Evolution is B*ad* S*cience*

T. Arthur Rasmussen

Evolution is B*ad* S*cience*

by T. Arthur Rasmussen

EvolutionIsBS.com

Published by TD Arthur Publishing, Portland OR

0011000100100000010000110110111101110010011010
0101101110011101000110100001101001011000010110
1110011100110010000000110010001110100011010 1

Cover photo from Shutterstock.com
Cover design T. A. Rasmussen

International Standard Book No. 978-1-733-5393-0-2

CAGCCTATCGACCGGCCGACCGGA
AGAAATCAATCAATGGATAGATCA

Pages 52, 181, 185 ACGT 00011011 KJV

Contents

About the Author *i*
Preface *v*

Part 1: In the Beginning

Is Evolution Possible? 23

Typing Monkeys 27

Computer Zeros and Ones 33

DNA 39

Similarities of DNA and Computers 45

What are the Odds? 51

Meteors and Mud Puddles 57

Microscopic Life 65

Part II: General Anatomy

From One Cell to What? 73

Body Parts 77

The First Heart 83

The First Brain 89

The Immune System 95

The Senses 101

The First Bone 109

Part III: Here and Now

The Fossil Record 117

Carbon-14 Dating 125

"Billion" Year Old Rocks 131

Viral Markers 139

Genetic Engineering 145

Control Systems 153

Big Bangs 159

Conclusion 167

Appendices

Appendix 1 173

Appendix 2 175

Appendix 3 177

Appendix 4 181

Cracking the Codes 183

Closing Words 187

To my wife Dianne

ABOUT THE AUTHOR

T. Arthur Rasmussen is an electrical engineer who has written a significant number of highly technical publications, specifications and industry standards. This experience would typically include making extremely complicated subjects understandable to his peers, and to the public and governmental agencies as well.

By chance, he became interested in the evolution debate and applied the same reasoning processes to the subject. His knowledge of computer programming lent itself well to analyzing the genetic programming found in all living things, as well as evaluating the incredible complexity of the designs found in the human brain. His background in electrical circuits furnished comparisons to nervous system networks, while his familiarity with electronics provided a background for analyzing fossil-dating machines.

He also reluctantly admits that a brief winning streak at a dice table as a young man led him to become very interested in probabilities, a science that happens to be extremely important when analyzing potential results of random genetic mutations.

In retrospect, he has also come to believe that this book did not come about by random chance.

I know this world is ruled by infinite intelligence.
Everything that surrounds us – everything that exists –
proves that there are infinite laws behind it.
There can be no denying this fact.
It is mathematical in its precision.

– Thomas Edison

A young man who wishes to remain a sound atheist
cannot be too careful of his reading.

– C.S. Lewis

This most beautiful system of the sun, planets,
and comets, could only proceed from the
counsel and dominion of an intelligent
and powerful Being.

– *Isaac Newton*

PREFACE

It's what you learn after you know it all that counts the most.
– John Wooden

I grew up out in the country, in northern Wisconsin. It was a wonderful childhood, with woods and fields and a nearby lake to explore with my siblings and our friends. There was never a dull moment, it seemed. Well, except for trips to town on Saturday for catechism and on Sunday for church. I didn't know any of the kids in catechism, and one of them was really annoying, constantly pestering me. Then, as if that weren't bad enough, on Sunday during church service the priest would start preaching in some strange language, and I'd always feel nauseated during the repeated and lengthy kneeling sessions. I was just a grade-schooler and didn't see the purpose in any of it. I wanted to be outdoors playing and enjoying nature instead.

Years later as a teenager, I decided I didn't want anything more to do with church or religion. I had lost all interest – mostly because I was being taught about evolution in school. That was all I needed for an excuse to convert to atheism at a fairly young age. I was quite certain about my newfound beliefs, and remained a know-it-all atheist until my mid-thirties.

That was when some of my religious co-workers began offering me creation-based reading materials, hoping to help me find my way. But I saw it all as just some old stories, and didn't believe any of it. I wasn't swayed at all.

It *did* start me thinking, however. What if I were to write some sort of anti-evolution book for those fellow employees, but with a strictly science-based approach? I might be able to sell a few copies and make some extra spending money. This made perfect sense to an atheist engineer like me. Even though I totally believed in evolution, I thought there must be some aspects of creation I could embellish by throwing together a few random facts and some scientific-sounding words.

Of course, this was all easily rationalized for an atheist. After all, it wasn't as if I planned on robbing a bank. I simply wanted to give religious people the kind of information I thought they were looking for. I truly believed I had everyone's best interests in mind.

So I started scribbling ideas in a notebook, one concept after another as they occurred to me, putting ideas together one piece after another. Then one weekend as I was busily working on my phony baloney book, everything strangely started to make sense. This made me more than a little nervous, so I stuffed the notebook into a drawer, thinking I'd just forget the whole thing.

For the next several weeks I managed to stay distracted with other activities. But then one day I found myself back at my desk, rearranging papers and puttering around. I kept noticing the drawer, knowing what it held, and I didn't want to open it. I really didn't. It kept nagging at me, however, and before long I reached down and carefully pulled the drawer open anyway. And there sat the notebook, its dark green cover staring up at me. I slowly lifted it from its resting place and set it on top of my desk.

I flipped the cover open and started turning the pages, leafing through, scanning my own arguments. Curiously, one section in particular held my attention. I read it slowly, carefully, and then all of a sudden a snapshot of total clarity snapped together in my mind – evolution really *is* impossible!

The swiftness of the revelation left me feeling dumbfounded, as if I had just been ambushed by an unseen force. I stepped outside for some air, hoping it would help clear the fog from my head. But all I could do was stare blankly at the sky and the trees, feeling confused – and even a little scared. Although everything around me was exactly the same, my world had just been turned completely upside down.

For the next several weeks, I went through some significant soul searching, which is highly unusual for an engineer. Eventually, I realized there was no going back. I would continue looking into the evolution story, come what may.

As I resumed my investigation, additional ideas occurred to me, one after another, and all were surprisingly *unfavorable* to the evolution side. I looked into every aspect there was, anticipating that I might have to concede random points here and there, but everything about the evolution story turned out to be lacking in overall logical analysis.

During this process, it also occurred to me that I should check the chapter about computers, to make sure my computer code examples didn't accidentally spell out any curse words. That would not be good. But a quick check verified there were no inappropriate words anywhere. So I had a good chuckle at my own expense, since the odds of bad words occurring randomly were incredibly small.

As I sat smiling about my irrational uncertainty, an idea came to mind. Perhaps I could *intentionally* make the computer code represent something in particular, and purposely hide a message inside this book. In fact, maybe I could hide all kinds of messages! Hey, why not?

And the coup de grâce? If someone didn't like the coded messages and wanted to ban my book from public schools, then the teaching of evolution, which proposes that all genetic code is accidental, would also have to be banned. In other words, if the extremely short coded messages hidden in this book could not have occurred randomly (as you will soon see) then there is absolutely no way that *3 billion* bits of genetic code in human DNA could have occurred accidentally!

As I devoted more and more time to analyzing the evolution story, I eventually realized that my writings had the potential makings of a serious anti-evolution book. But it wouldn't be the kind I had first envisioned, not even close.

It ended up taking many years of research and analysis, but eventually I settled on this version, which is intended to be introductory and not overly complicated. It turns out, simplicity is not easily achieved with such a wide-ranging subject, but this is an attempt in that direction.

Also, during this entire process, I went through a slow-but-steady personal transformation, totally rejecting atheism in favor of something I found to be far more logical – and which I now firmly believe is the ultimate intellectual pursuit.

No matter where you currently are in your own walk through life, I sincerely hope you find this book to be interesting, informative, and maybe even life-changing. It certainly has been for me, every step of the way.

PART I

In the Beginning

Chapter One

IS EVOLUTION POSSIBLE?

One

IS EVOLUTION POSSIBLE?

Whenever anyone says 'Theoretically,'
they really mean 'Not really.'

– Dave Parnas, software engineer

If evolution were possible, it would take *billions* of miracles. Generally speaking, a lifeless planet would have to genetically engineer a worldwide ecosystem into existence – while utilizing no intelligent input whatsoever. As we will see, the odds of that ever happening are zero. Even the very simplest microscopic species contain genetic code that is too incredibly complex and intricate to have had any chance of evolving accidentally one single species at a time into an ultra-complicated system.

The entire evolution story completely ignores anything and everything involving probabilities. When backed into a corner, many evolutionists will eventually admit that their beliefs go completely against the odds. But they just shrug it off, insisting that "everybody knows" evolution is factual.

In reality, the evolution story has a *zero percent* chance of happening, whether mathematically, logically, biologically, or via any scientific discipline, as this book will show.

So, to set the stage for this investigation, let's begin by looking at whether the odds of any particular thing happening can actually go to zero, as peculiar as that concept might seem.

We can start with an ordinary example. Suppose you held a golf ball out at arm's length and then dropped it onto a concrete surface. The ball would bounce about half-way up from where you originally dropped it, and with each successive bounce it would reach about half the height of the previous bounce. So, what are the odds that the golf ball would continue to bounce for one full day? How about for one full hour? OK, how about for just one full minute?

In reality, the golf ball would very quickly settle down and completely stop moving, typically in under ten seconds. Each bounce would be half as high as the previous bounce, and would quickly drop to exactly zero.

On our level of existence, physical differences can only get so small and then that's it, they go to zero. This is plain to see with the smaller and smaller bounces of the golf ball, until it no longer moves at all.

As you will see in the examples used throughout this book, one chapter after another, and one aspect after another, the possibilities of evolution actually happening are zero. This is particularly true regarding DNA, the incredibly complex building block that creates all life on Earth. The "golf ball" odds for *accidental* genetic code bounce lower and lower, quickly bringing the evolution story's probabilities down to zero, where they stay – permanently.

Chapter Two

TYPING MONKEYS

Two

TYPING MONKEYS

The most exciting phrase to hear in science,
the one that heralds new discoveries,
is not 'Eureka!' but 'That's funny...'
– Isaac Asimov

Not that long ago, evolution textbooks would use a make-believe example that had a typing monkey (you read that right – a typing monkey!) produce an exact copy of a Shakespearean play completely by accident. Supposedly, this same monkey then went on to accidentally produce copies of more plays, and even managed to produce novels and poetry purely by random chance.

Naturally, the typing-monkey story presumed that the primate took an extremely long time to achieve its very lofty goals. But this raises an important question. Just how long would it actually take a monkey to type out an entire play accidentally? Better yet, how many monkeys would be needed to type out an entire play in any given amount of time? Let's investigate.

First, we'll look at the odds of the typing-monkey story actually happening. We'll say that one lone monkey sits at a typewriter for one hour per day, eagerly tapping at the keys, at a rate of two keystrokes per second. As a target for the monkey to type, we can use "ooimcmooomoixixiiixiicmmooooimciixmicmimo" as a possible letter-combination goal. All the monkey needs to do is accidentally type those letters in that order. Not an entire play, mind you, just those letters as shown.

And let's make it as easy as possible for the monkey by greatly simplifying his typewriter. Rather than operating a standard model that has over 40 keys with letters, numbers, punctuation marks, shifting, and a spacebar, we'll have him use a make-believe typewriter that has just the lower case letters "a" through "z". This way, the monkey will only have to tap at a maximum of 26 keys, significantly increasing his chances for success.

With the stage set, the monkey then types for one hour a day for its entire 70-year lifetime. So, how well do you suppose the lone monkey will do? As it turns out, despite all that effort, the monkey would only get the first five letters, "ooimc," of our target letters "ooimcmooomoixixiiixiicmmooooimciixmicmimo."

Some people who read the above statement will have a very difficult time believing it is accurate. If you have trouble moving forward after this first bombshell, see Appendix 1 for confirmation of the technical details. Then, mentally prepare for the onslaught to continue.

OK, now we'll improve the odds and try a scenario that uses ten monkeys – yes, ten monkeys – each hitting four keys per second instead of just two keys per second. Let's also have these ten monkeys type for ten hours per day, instead of for only one hour per day. How well do you think ten monkeys typing ten hours per day would do?

Ultimately, the best this team of primates could do over a 70-year period would be just the first seven letters, "ooimcmo," of "ooimcmooomoixixiiixiicmmooooimciixmicmimo." While just one monkey typing two keystrokes per second for one hour per day would get only five letters, ten monkeys typing four keystrokes per second for ten hours per day would achieve only seven letters. And this is with the greatly increased primate-power! (Technical details are in Appendix 1 for all non-believers.) Quite clearly, the monkey story is rapidly dying on the vine.

So now, just for fun, let's really step things up. According to some unproven stories, our universe is said to be several billion years old. So let's cover every unsubstantiated guess out there and have *one million* monkeys type 100 keystrokes per second, 24 hours per day, for *100 billion* years. Mind boggling, yes, but these are miracle monkeys burning up the keyboards.

So, just to be clear, we have one million monkeys typing nonstop for 100 billion years. How many letters of "ooimcmooomoixixiiixiicmmooooimciixmicmimo" do you suppose this massive army of monkeys would type over the course of such an unimaginably long period of time?

Amazingly, the million monkeys would only get 18 letters, "ooimcmooomoixixiii," – and that's it! Just 18 letters after typing at ultra-high speed for 100 billion years! Unbelievable, isn't it? Well, not really. Again, anyone doubting the science behind these figures is referred to Appendix 1, or simply do an internet search on the power of geometric progressions.

The odds of the million monkeys succeeding in their quest to type out all 41 letters of "ooimcmooomoixixiiixiicmmooooimciixmicmimo" quickly go to zero and stay at zero permanently, just like the bouncing golf ball. But what else should we expect from a story that refers to monkeys sitting at typewriters, trying to accidentally type out literary classics?

The odds would be exactly the same for anything the million monkeys were assigned to type. They would only manage to achieve the first eighteen letters of any particular work. Give them the task of typing Hamlet, and they would only manage to type out "*elsinoreaplatformb,*" from Hamlet's "*Elsinore. A platform before the castle,*" and that would be it. Eighteen letters and not one letter more. And all of this is done with one million monkeys using our modified, monkey-friendly typewriters!

How long would it take one single monkey to type out an entire play using a real typewriter? Forever and a day plus several eternities. The typing-monkey story uses zero common sense to begin with, and is utter nonsense in the end.

And while we're on the subject of randomly occurring letters, how likely would it be for all seven of the letters abutting the left margin of the first paragraph of Chapter One to randomly spell a one-word message? As it works out, the probabilities against accidentally getting those seven letters in that exact order are about *8 billion* to one. So, could that hidden message have occurred accidentally? Not a chance in the world.

See Appendix 1 for this Chapter's technical conclusions.

Chapter Three

COMPUTER ZEROS AND ONES

Three

COMPUTER ZEROS AND ONES

In our profession, precision and perfection are not
a dispensable luxury, but a simple necessity.

– Niklaus Wirth, computer scientist

Computer software is designed using precisely arranged codes that produce programs to determine what a computer can do and what its functions will be. If one single digit of code is missing, switched, or out of place, the computer might not function properly. Any software developer will confirm the frustrations of this reality.

For example, there was tremendous concern surrounding the so-called Y2K situation as the end of year 1999 approached and 2000 neared. It was feared that computers might not be able to handle the switch from the "99" in 1999 to the "00" in 2000. Those concerns turned out to be excessive, in part because the situation was addressed ahead of time. Nevertheless, it did illustrate the very real concern regarding computer code's sensitivity, since even a very small error could result in serious problems.

Considering the complexity and preciseness required, would it be possible for a computer program to happen on its own? This undoubtedly seems like a ridiculous proposition. How could something so amazingly complex happen by chance? But since the evolution story says that genetic code came about randomly, let's examine computer code and the probabilities against it occurring accidentally. (This will be a warm-up for Chapter Four where we will investigate the real deal – DNA).

Computer software uses a code system that is made of 0's and 1's arranged into segments only eight digits long. For example, the letter "A" is represented by 01000001. The letter "B" is represented by 01000010. The letter "C" is 01000011, and so on for the entire alphabet of A through Z.

At first glance, it might seem that there wouldn't be enough spaces available to encode every single letter of the alphabet using a mere eight 0's and 1's per letter. But surprisingly, with just eight positions available, there are actually 256 different combinations of 0's and 1's possible. Those 256 variations leave plenty of room for each of the 26 small letters of the alphabet, the 26 capital letters, the numbers 0 through 9, all of the punctuation marks, all of the symbols, and so on, with room to spare. When a computer sees a string of 0's and 1's eight digits long, it "knows" exactly what letter, number, punctuation mark or symbol that particular string signifies.

Now, to start our analysis as to whether computer code can happen accidentally, let's take a common coin and assign heads as 0 and tails as 1. We can flip the coin, keeping track of the results. Could a random coin-toss list of 0's and 1's end up representing an actual computer program? Let's find out.

We'll start by trying to get just the first eight 0's and 1's of a randomly chosen program. The coin would be tossed, and since there are only two possible outcomes – heads or tails – we would have a successful toss once every two tries, on average. However, every time there is an incorrect toss, we would have to start over, since the program could be corrupted by an incorrect digit.

As it turns out, in order to get just the first eight digits of the computer program correct, we would need to overcome odds of two times itself a total of eight times, which multiplies out to 256. That means we would have to toss the coin 256 times on average just to achieve a rather meager eight computer digits – which would be only one letter of the alphabet!

To accidentally get 24 of the computer 0's and 1's lined up in the correct order, such as with a randomly chosen 0100011101 10111101100100 (which would only be three letters of the alphabet), you would need to overcome odds of multiplying two times two a total of 24 times – which would be 16,777,216 tosses of the coin! And that is for only three letters of code! We have essentially reached zero chance of getting things correct since no one would toss a coin over 16 million times just to get three letters of the alphabet!

As we have already seen, multiplying over and over again ultimately produces extremely large numbers, which holds true in this case. To accidentally get 144 computer 0's and 1's lined up exactly in the correct order, such as with an arrangement of 0101000001110011011000010110110001101101001000000110 0110011010010110011001110100011110010110111101101110 0110010100100000011100110110100101111000 (which would be only 18 letters of the alphabet, including spaces), you would need to overcome odds of multiplying two times two a total of 144 times, which would be somewhere in the neighborhood of 22,300,000,000,000,000,000,000,000,000,000,000,000,000,000 flips of the coin! It would take a ridiculously massive number of coin tosses just to get the first 144 of the 0's and 1's in the correct order (which would produce only 18 letters of the alphabet). Going beyond 144 digits, the numbers grow even more gargantuan against software happening by accident.

This explains why intelligent computer programmers are required instead of random generation methods. There are far too many possibilities to ever produce more than a tiny fraction of any computer program randomly – even at computer speeds. Additionally, the results could never all be run through a computer one by one and observed to see what they could do; there are too many possibilities and the time involved would make the task impossible. Furthermore, since every program works in conjunction with other programs, every potential program would have to be run through a computer not just once, but a near-infinite number of times. Then a near-infinite number of results would have to be observed and analyzed. Such things could never happen.

Also, as food for thought, if in an attempt to write Hamlet, one million computers compiled one hundred keystrokes per second for 100 billion years, they would not even get the first four full words. And yes, if that sounds familiar, the probabilities are exactly the same as for the typing monkeys in Chapter Two.

Ultimately, there is no way that programming such as 01001 0010111001101100001011010010110000101101000001000000 0110100001110000011101000110001001110011 could occur accidentally. And the odds against an entire computer program happening by random chance are basically equivalent to zero, just like in the golf ball example in Chapter One.

Chapter Four

DNA

Four

DNA

We've discovered the secret of life.

– Francis Crick

Note: This chapter can get a little complicated, because DNA is so incredibly complex. Feel free to just skim over this information, but bear in mind the intricacies of the codes and mechanisms involved.

In the previous chapter we discussed how 0's and 1's form computer programs. In a similar way, genetic code uses four molecules to create all life on Earth. Yes, just four different molecules can be arranged in specific orders to produce living things. These four miraculous molecules are adenine, cytosine, guanine, and thymine. They are referred to by their initials A, C, G, and T, and are also categorized as being amines.

A, C, G, and T form much larger molecules called DNA (which is short for deoxyribonucleic acid.) All DNA is comprised of these four amines lined up in a very specific order, much like the digits in a computer program. Every living thing on the planet is created via its unique genetic code. This means that DNA alone is capable of producing anything from tiny bacteria all the way up to a blue whale.

Also, there is another type of genetic code, RNA, which stands for ribonucleic acid. RNA uses U (uridine) rather than T (thymine). So RNA's amines are A, C, G, U, very much like the

A, C, G, T that DNA uses. DNA is used in some areas of a cell, and RNA is used in others.

This completes the first part of the genetic code lesson. Are you ready for part two?

All amines (A, C, G, T/U) are arranged into items called codons. A codon is simply three amines in a row, such as ACG or CAC. So, starting with AAA and going through TTT (or through UUU), there are 64 different combinations possible. Therefore, 64 different codons are possible.

Now we come to something that is truly remarkable. Every cell in every living thing has several very complex mechanisms in place to "read" these codons. The codons direct the specific construction of 20 different amino acids, which then combine to produce proteins, which then form the anatomical structures and specific organs of all living organisms. All of this happens per the DNA's specific instructions.

For example, if a cell sees the codon CAA, it will produce the amino acid glutamine. If it sees the codon AGC, it produces the amino acid serine. Even more incredible is the fact that the codon UGA is recognized by a cell as the "Stop" codon, signaling the end point of each individual construction chain. Furthermore, the codon AUG typically dictates the "Start" point of construction.

This brings us to another interesting aspect. Each of the 64 codons specifies one of the 20 amino acids or "Start" or "Stop." Therefore, most of the amino acids have two or more codons that will each produce them. For instance, CAU will instruct a cell to produce the amino acid histidine, as will CAC. As another example, UUA will produce the amino acid leucine, as will UUG, or CUU, or CUC, or CUA, or CUG.

It is essential that amines and the resulting codons be correctly arranged. Unfortunately, in the vast majority of cases, if any amines are out of order, the codons will be changed, and infirmities and genetic diseases can occur. As an example, hemophilia is caused by the reversal of a single amine pair. What should be a T-A amine combination is unfortunately an A-T, which causes the hemophiliac's blood to not clot properly. Sickle cell anemia also is brought about by the reversal of a single T-A amine pair. These are only two examples that illustrate the precise amine alignment required in DNA.

There are thousands of genetic disorders that have been identified to date. This illustrates the fact that if DNA is randomly altered, the codons are altered, and the associated amino acids are negatively changed in the organism.

OK, that completes our opening lesson on DNA. Using this information as a basic foundation, we will continue our discussion in upcoming chapters.

Chapter Five

SIMILARITIES OF DNA AND COMPUTERS

Five

SIMILARITIES OF DNA AND COMPUTERS

When I see a bird that walks like a duck
and swims like a duck and quacks like a duck,
I call that bird a duck.

– James Whitcomb Riley

In previous chapters, we learned about computer code and genetic code. Amazingly, these two entirely different programming systems have incredible similarities, such as the following:

Computers use 0's and 1's to form computer code.
DNA uses A's, C's, G's and T's to form genetic code.

Computers use 8-digit arrangements of 0's and 1's to form letters, numbers, punctuation marks, and other items.
DNA uses 3-amine arrangements of A's, C's, G's, and T's to form codons and start/stop points.

Computer code is held on microchips.
Genetic code is held on genes.

Microchips are interconnected on circuit boards.
Genes are interconnected on chromosomes.

Circuit boards communicate back and forth with each other.
Chromosomes communicate back and forth with each other.

A computer runs its programs through a microprocessor.
DNA runs its programs through a ribosome.

Computers use a variety of routine subprograms.
DNA uses a variety of routine subprograms.

A computer produces complex outputs via its programs.
DNA produces extremely complex outputs via its programs.

Computers contain self-correcting software.
Cells contain intricate self-correcting DNA.

Computer programs are very long, complicated sets of instructions, and must be perfect and without interruption. For example, a randomly chosen selection of zeroes and ones such as 00.10..110.111.100..011.0..1.010.001.000.1 would not work in a computer because of the periods scattered randomly throughout. Anything other than zeroes and ones in a very precise order could cause a computer to immediately crash. Similarly, DNA must use only the amines A, C, G, and T, with no other amines allowed in the rungs of the double helix, and RNA must use only A, C, G, and U.

Typically, computer programs and their associated subprograms have so many 0's and 1's that printing them out on paper would require thousands of pages. For example, inserting a program with 10 million bits of computer code into the end of this book would add a very significant 2,500 pages.

By comparison, human genetic code contains about 3 billion bits of information in the form of amines. If this were tacked onto the end of this book, it would add 750,000 pages, which would produce a book 250 feet tall – as high as a 25-story building!

Lastly, consider this. The earliest computers were huge, and filled an entire room with their many tubes and old-fashioned electrical devices. But despite their large size, they were not very fast, nor were they very powerful.

Over time, the more advanced mainframe computer was developed. Although being somewhat less massive in size, it was capable of doing more things faster. Eventually, the desktop computer was developed, followed by the laptop and then the handheld, getting smaller and yet more powerful with each advancement. However, thousands of years before manmade computers were invented, there was the microscopic cell and its incredibly complex DNA – the smallest and most advanced computer system of all time.

Modern-day software programs can't compare to DNA's vast life-producing complexity. All genetic code, regardless of how unappreciated it might be, is more amazing than any software known to man, by far. Arranging a huge amount of information exactly as needed in a microscopic space, with each component that cells require to process DNA also in place, cannot occur by evolution's storyline in any way, shape, or form. Every 0 and 1 needs to be precisely ordered, as do every A, C, G, and T.

Chapter Six

WHAT ARE THE ODDS?

Six

WHAT ARE THE ODDS?

We will now announce the winner of the lottery,
followed by the names of the 4,347,608 losers.

– Sidney Harris
(in comic strip, "Explaining the Odds.")

Let's begin our investigation into whether DNA can happen accidentally by taking an up-close look at an E. coli bacteria cell, which has the smallest known genetic code of any cellular life form. Despite ranking at the very bottom on the cellular DNA scale, an E. coli bacteria's genetic code just happens to contain about 4,600,000 amines. That's right – there are roughly *4.6 million* amines in this particular microscopic bacteria's DNA. These amines are arranged to form genes, and the genes are all nicely packaged into one chromosome.

To illustrate the magnitude of E. coli's DNA, if we were to print out its 4.6 million amines on paper, with all the amines represented as A, C, G, and T, and in the font size used throughout this book, we would need an amazing 1,150 pages! That's a very significant amount of DNA, especially for a so-called "simplest" single-cell life-form. With this information in mind, what do you suppose the odds are of bacterial DNA happening by accident? Let's find out.

For starters, we can use a very small example to illustrate the odds against accidental DNA formation. We'll use just the first

24 amine positions of a randomly chosen alignment, such as CCAACTATCGACCGTACGTCATAC. As we already know from Chapter Four, each codon is three amines long, so the resulting 8-codon sequence would be (CCA)(ACT)(ATC)(GAC)(CGT)(ACG)(TCA)(TAC). Since some amino acids can be coded by up to six different codons, we'll use the best-case odds, ignoring cellular bias toward particular codons and the "Start" codon, just to simplify things.

So, what are the odds of those particular eight codons lining up correctly? Brace yourself! We would need a string well over *91 billion* codons long. Then somewhere in that huge mix would be the extremely tiny 8-codon run. Yes, that's correct. Approximately 91,625,967,549 codons would be needed to get only eight codons in the correct order! (See Appendix 2 for additional technical details.)

Now, to bump things up just a little bit, instead of only eight codons, we can determine what it would take to accidentally get 192 codons correct, which would be 576 amines, such as shown in the arrangement below:

AGAACACGCGTTCTAGAGAACACTCGTTCGCAAGAACTATCGTTAGAACGTA
CGTTCTCGCGCCCGCAAGAACTCACGGACGCCAGAACTCTCGTTCTAGCGTA
CGCAAGTAAGAACTCACGGACGACCTCAAGAACGGACGCCAGAACGCTCGAC
CTCGCGCCAGAACGGACGGCCTATAGAACGTTCGTGCGTACTGCAGAACGAG
CGCCCGCTCGTTCTCACTCACGCCCGTGAGAACCATCGTTCGTGAGTAAGAA
CTCACGGACGACCTCAAGAACTCTCGGACGTTCTATCGTTCGCCCTCGCGCC
CTAGAGAACGAGCGCCCGTACGGCCGCCCTCGCGCCCTCACGGAAGAACGGC
CGTGAGAACGGACGGCCGTCAGAACTATCGGACGTTCTCCCGTACGCAAGAA
CGTGCGTTCTCAAGAACTAACGCCCTAGCGGCCTATCGGAAGTAAGAACGAG
CTCCCTCAAGAACGGACGACCTCGCGCCAGAACGCCCTCGCGCCCTAGCGTA
CGACCTATCTCACGGCCGTGCGCTAGAACGTACGGCCGCGCGCCAGTGAGAA
AGAA.

The odds against getting those 192 codons lined up perfectly calculate out to a number so enormous it seems impossible – because it is impossible! The odds against this happening are roughly 772,020,010,011,010,110,000,101,110,100,011,101,000,110,100,001,100,101,011,101,110,010,000,000,100,000,011,011,100,110,100,101,101,110,011,001,010,111,010,001,100,101,011,001,010,110,111,000,100,000,001,000,000,111,010,001,110,111,011,001,010,110,111,001,110,100,011,110,010,010,000,001,110,011,011,010,010,111.1000 to one. And this is simply to get a tiny percentage (0.0125%) of the 4,600,000 amines needed to create one "simple" E. coli bacteria cell, leaving 4,599,424 amines still to go.

With this simple illustration alone, it should be obvious that DNA of *any* kind could never happen on its own. But if you're still not convinced, take a moment to consider how much more happens *within* DNA. Remarkably, some A's, C's, G's, and T's pull double-duty, not only creating amino acids, but also simultaneously functioning as genetic regulators, enhancers, and promoters. This means that a significant number of amines defy astronomical odds to not only create amino acids, but also simultaneously participate in gene control. Yet another amazing aspect of DNA is that there are markers attached to genetic code that can turn genes on or off as needed to adapt to their local environment. Incredible!

Could all of those complex mechanisms and their associated programming have happened by accident? Not a chance! If you're still in doubt, the next chapter provides even more substantiating evidence.

See Appendix 2 for this chapter's technical conclusions.

Chapter Seven

METEORS AND MUD PUDDLES

Seven

METEORS AND MUD PUDDLES

Nothing in life is to be feared, it is only to be understood. Now is the time to understand more, so we can fear less.

– Marie Curie

The evolution story postulates that long, long ago, meteors entered Earth's atmosphere and spread amines into the air. Supposedly, those amines randomly drifted together to form the world's first-ever genetic code. Allegedly, this original DNA then incredibly combined with various items that just happen to create cells, such as a cell wall, a ribosome, cytoplasm, etc. Evolutionists believe that those individual cell parts not only appeared all on their own – but that they also somehow magically appeared in the very same microscopic location as the DNA.

However, in order for that part of the story to be even remotely workable, something miraculous would have to somehow coincidentally happen. The randomly occurring DNA would have to accidentally create *every one* of those various cell parts – which supposedly had already occurred all on their own! So, the accidental DNA that creates cell parts came along after the cell parts came along by accident? That is a very strange concept!

Could such a miraculous event actually occur? Could meteors dump a sufficient number of amines into the atmosphere to accidentally create DNA? Then, could a multitude of cell parts accidentally encoded by that DNA be accidentally abounding in the same location? Absolutely not.

So-called "simple" life-forms are comprised of a single cell, but are nowhere close to being simple. As explained in the previous chapter, E. coli bacteria have somewhere in the neighborhood of 4,300 genes, and each gene has on average about 1,070 amines, for a total of approximately 4,600,000 amines in all.

To begin this investigation, we can use an example similar to Chapter Two's one million monkeys, who typed 100 keystrokes per second for 100 billion years and only succeeded in getting the first three words of Hamlet. In this case, we will be looking at the likelihood of accidental DNA formation.

We can start by pretending that 10,000 meteors entered Earth's atmosphere every single day for one million years. In addition, we'll say each of these meteors disbursed 10,000 amines into the air, and all of those amines joined together to form long strands. We can then analyze the total number of amine combinations possible, and from that, determine what would be the longest arrangement of correctly aligned codons.

If a bacteria cell's 4.6 million amines were printed out on paper, it would fill 1,150 pages. Taking that into consideration, how much bacterial DNA do you think 10,000 meteors entering Earth's atmosphere every day for one million years might produce? Incredibly, there would be no more than 36 amines lined up correctly! Once again, the result is shockingly small, but the numbers speak for themselves (see Appendix 3 for details). Those 36 amines would not even take up the first line on the first page of our stack of paper, leaving the rest of the 1,150 pages completely blank! The genetic code necessary to create a basic life-form would not have even barely begun, let alone be anywhere near complete.

Of course, evolutionists might claim the meteor bombardment could have been even more severe. We can accommodate their claims – and then some. Let's pretend that meteors streaked into Earth's atmosphere at the rate of one every single second. This would total 86,400 meteors per day. Let's also pretend that this continued non-stop for 3 billion years! Even more, let's say each meteor carried 10 million amines that disbursed into the air, forming long strands. Amazingly, this exaggerated, impossible scenario would only produce a meager 54 amines lined up in the correct order – leaving about 4,599,946 million amines still to go. This means that the odds of meteors creating life essentially equal zero. (See Appendix 3 if you don't believe it!)

As we know, there is another scenario proposed by the evolution story, which is that life might have somehow started in a primordial soup. What are the odds that a mud puddle gave rise to life on Earth? It should not be surprising to learn that the answer is zero chance.

To cover all primordial soup possibilities, let's first choose a very special mud puddle comprised of only the amines A, C, G, and T – with no water, silt or anything else – just a large pool of amines. In addition, let's say the pool was actually the size of an ocean, one thousand miles long, one thousand miles wide, and ten miles deep, which would be 10 million cubic miles of amines all lined up in long strands. So, how much DNA do you think such an amine ocean would create? Incredibly, it would produce a string only 105 amines long. That's right, once again it is several million amines short of anything genetically meaningful! (See Appendix 3 if you doubt the figures!)

Lastly, there is some speculation that amines might shuffle and rearrange themselves over extremely long periods of time. Let's put that notion to rest by pretending that at one time the entire planet Earth was 100% amines A, C, G, and T. We'll even go beyond ridiculous and say that all those amines shuffled 10 times per second for 100 billion years. Are you ready for another shocker? Astoundingly, that scenario would produce a string only 159 amines long! Once again about 4.6 million amines short of accidentally creating DNA. (See Appendix 3.)

In reality, a few amines scattered throughout mud puddles here and there would be lucky to achieve 15 amines lined up correctly, which is just a tad bit short of the 4,600,000 needed for a single bacteria cell.

Nonetheless, most evolutionists insist that a very short string of 15 amines could have added one amine at a time and built up the correct sequence while maintaining the prior order, steadily increasing in length, eventually reaching the properly ordered 4,600,000 million amines.

That entire concept is absolutely ridiculous and is essentially a "pick-and-choose" method. This is especially true when starting from nothing. A source of intelligence would be needed to determine which amines to keep and which ones to throw away. There isn't anything even remotely random about the pick-and-choose method since there has to be outside input from a knowledgeable entity or it simply could not happen.

Consider a bank vault with a combination lock that can only be opened by spinning a dial back and forth to four pre-selected numbers. Starting with the first number, there are only two ways to find out what the first number is; you can either ask a source of intelligence what the first number is, or you can try to randomly guess it. But you cannot just guess the first number, and once you have it, guess the second number, and once you have that, guess the third number, and then guess the fourth number and open the safe. It doesn't work that way. You need to guess all four numbers at the same time. You wouldn't know what the correct first number was until you had guessed all four numbers correctly. That is why safes are extremely safe. There are so many possible combinations that guessing all four numbers at the same time would take tens of millions of attempts! And of course, as you might have guessed, the same reasoning applies to the appearance of DNA's amines.

Chapter Three showed how computer programs cannot be generated randomly, since accidentally hitting upon all the necessary zeroes and ones at the same time would be impossible. It is the same concept for genetic code. It is simply not possible for *4.6 million* bits of DNA to accidentally hit upon the right combination for each gene, all at the same time, with no one there to meticulously put the amines into the necessary order, and no one keeping the correct amines and discarding the incorrect ones. Not a chance.

Probabilities prove that if mutations were actually random, then almost all of them would be detrimental. This also means that beneficial mutations are close to impossible. If truly random mutations were to continually occur, all life would soon vanish. Allowable variations within species are not accidental – they are specifically designed to be of potential benefit depending on environmental circumstances.

Consider the Empire State Building. It was not designed and constructed by accident. Yet we can make some minor changes to it, such as painting one of its rooms green, changing the window blinds in another room, and changing the carpets in an office suite. But being able to make very minor changes does not mean the Empire State Building evolved from nothing.

Here is another point of interest. When an animal dies and its body starts to decay, its DNA starts to decay also. That's why scientists have not been able to clone extinct species such as the carrier pigeon or the wooly mammoth. The available DNA has deteriorated too much, and it will continue to decay more and more as time goes by.

Of additional interest, consider that "dead" DNA does not construct new DNA, even though all of the necessary components are right there. That simply does not happen. So, what chance would extremely tiny amounts of DNA randomly scattered in the air or in water have of accidentally creating anything useful? There is zero chance.

It's time for evolutionists to give up on their meteor and mud puddle stories. They are stories and nothing more. Genetic code cannot happen on its own. It is far too complicated and massively long to have lined up perfectly out of total randomness.

See Appendix 3 for this Chapter's technical conclusions.

Chapter Eight

MICROSCOPIC LIFE

Eight

MICROSCOPIC LIFE

The proper study of mankind is the science of design.
– Herbert Simon

Viruses

Viruses are incredibly small. They are so tiny, there is disagreement about whether they even qualify as being life forms. Viruses don't have a cell wall or any of the internal parts that a living cell has. In fact, they make bacteria cells look huge by comparison. Yet despite their tiny, tiny size, viruses are extremely complex. They contain genetic code, are comprised of proteins, and come in a variety of forms. Unfortunately, viruses can perform some truly dastardly feats.

Typically, when a virus finds a living cell, it latches onto its surface. Next, the virus inserts its stinger through the cell wall and injects a copy of its own DNA into the interior of the cell. The unfortunate victim cell then "reads" the alien DNA and starts producing virus after virus. Within a short time, the cell becomes overstuffed with viruses, causing it to burst. This releases all of those brand new viruses into the host organism. Then, all the new viruses find host cells of their own and repeat the entire process, rapidly reproducing. The viruses continue to swiftly multiply until the host organism either mounts a successful defense or dies from the onslaught.

There happens to be a particular kind of virus called a bacteriophage whose head is made of a protein which contains DNA. The DNA-filled head is connected to a collar, which in turn is connected to legs and a tail. This combination of head, tail, and legs somewhat resembles a microscopic spaceship! Its legs are used as "landing gear" when the virus attaches to the surface of a host cell. The virus then goes about its business of penetrating the cell wall and injecting its DNA, setting its own reproduction into motion.

With all that in mind, let's now consider the hepatitis D virus. As you remember, a virus is considerably smaller than a bacteria cell. As a matter of fact, the hepatitis D virus happens to have the *smallest* known genetic code of all, with only 1,700 amines. Compare that to a microscopic E. coli bacteria's genetic code which has 4,600,000 amines! But, despite its simplicity, could this "extremely simple" hepatitis D virus with its extremely small genetic code have come into existence by accident? No way.

As we learned in the previous chapter, a 15-amine lineup would be the theoretical accidental maximum in the real world. So in the case of the hepatitis D virus and its smallest known genetic code, that would leave over 99% of its amines still to go in order for this nasty little "bug" to create itself!

If individual virus parts *had* floated around in a mud puddle, such as the evolution story postulates, where did those parts come from? Why would they all drift to the very same microscopic space, and then interconnect for no particular reason? Better yet, how could the DNA that *creates* those virus parts also be in the same location at the same time, and just happen to float inside the magically formed head of the virus? The answer? It could only happen in a make-believe story.

If you're still not convinced, then consider this. *If* a virus had defied all odds and somehow miraculously came about all on its own as the very first life-form ever, then what would have been its purpose? There wouldn't have been any cells around for it to attack. It would have had no use for its invasive mechanisms. It would have served no purpose at all. And since there would have been no recipient to inject its DNA into, it could never have reproduced. Pity that poor first virus, all dressed up and absolutely nowhere to go.

Bacteria

Bacteria cells are at least ten times bigger than the largest virus in existence, and are far more complicated than any virus. Therefore, it should be obvious that bacteria have even less chance of occurring accidentally. But according to the evolution story, a bacteria cell's parts were randomly created by murky water, and were then perfectly assembled via some occasional sloshing action.

However, bacteria are comprised of a cell wall, plasma membrane, cytosol, flagella, DNA, ribosomes, and many other components, all with their own individual parts that are smaller yet. Not only that, but the bacteria's 4,600,000-amine instruction book of precise genetic code would have to have accidentally occurred in the very same puddle at the very same time, which is simply not possible.

The Ribosome

Consider how all living cells have a component called a ribosome, which utilizes DNA in a very complex way. First, the DNA is changed into messenger mRNA. This mRNA is then run through the ribosome, which uses pre-constructed sets of transfer tRNA containing specific amino acids that combine into proteins as directed by the particular genetic instructions. The proteins are then directed to form specific organic components. Wow! And that is the "condensed" description of a ribosome without the transformation details regarding anti-codons, codons, translation, start-codons, nuclei, initiation, nuclear envelopes, and much, much more.

As if that weren't enough, the amazing ribosome will also analyze RNA and DNA to determine whether they contain any errors. And as astounding as this may seem, if genetic irregularities or errors are found, the ribosome not only detects them – it corrects them!

The ribosome is just one of many cellular components that are so ultra-complex they have "designed" written all over them. Additionally, every component that works in cooperation with a ribosome had to have been individually created and then furnished with genetic code. Does this sound like a system that could evolve from a mud puddle? Definitely not!

PART II

General Anatomy

Chapter 9

FROM ONE CELL TO WHAT?

Nine

FROM ONE CELL TO WHAT?

Science is nothing but perception.

– Plato

What does the evolution family tree show as the next step up from microscopic single cells? For some reason, it shows fully formed plants and animals! But how could that be possible since single cells multiply by dividing into two separate single cells? When a bacteria cell splits, it always splits into two bacteria cells, never into one bacteria cell and one fully formed jellyfish.

How could the world's very first cell have gone on to divide into two cells completely by accident? To reproduce, a cell first makes a duplicate of its own DNA. It then creates copies of its other components. The cell wall then contracts inward, separating the two sets of DNA and the other items, eventually sealing itself off in the middle and forming two separate but basically identical cells. Needless to say, this is no easy task!

Could there have been a transitional phase from a bacteria cell into a multi-cell creature? No. But if you believe the evolution story, it must have been at least a two-cell creature. However, two cells stuck together would have had no possibility of creating an advanced life-form, and neither would any amount of bacteria cells randomly globbed together.

Of course, there are some single-cell creatures that are able to "glom onto" one another. For example, slime mold cells not only group together and form what looks like pond scum, but they also move around in the water as one unit. That really is pretty remarkable, and it would obviously require advanced programming for such cooperative interworking to be possible. However, a globule of slime mold cannot create a heart, liver, kidney, etc. It is simply a "community" of slime mold cells working together as a group in perfect unison.

Could the next step up from a single-cell organism have been a jellyfish? Absolutely not. For such a thing to happen, the DNA for a jellyfish would have to be created somewhere and then get injected into a lone cell. That cell would then have to go about the process of producing jellyfish parts per the DNA's directions, which would make it quickly explode. After all, how could a bacteria cell produce a multi-million-cell jellyfish, all while possessing nothing in the line of such reproductive capabilities?

As we know from earlier chapters, DNA cannot happen on its own. Further, when a virus injects its DNA into a cell, there is zero chance of that DNA randomly mutating into the DNA for a different creature. There is no way the genetic code necessary to direct the entire process could come into existence by chance.

A single-cell organism cannot happen accidentally. It can only make replicas of itself – or of an invading virus. Its DNA cannot accidentally mutate into different DNA that happens to produce a multi-cell organism. Not a chance in the world!

Chapter 10

BODY PARTS

Ten

BODY PARTS

A designer knows he has achieved perfection
not when there is nothing left to add, but
when there is nothing left to take away.
– Antoine de Saint-Exupery

A surprisingly large number of anatomical components exist in the human body. In a typical anatomy textbook, the index alone is usually well over 40 pages long, and that's just to list the individual parts by name. An anatomy textbook's full contents will contain over 1,000 pages, with highly detailed illustrations of the location, design, and function of each and every item in the human body.

The complexity of the human body cannot possibly be overstated. Textbooks often describe it as being an inconceivably complex machine, having a highly complicated design. Listed below is a small sampling of some of the human body's various components:

- Red blood cells
- Optic nerve
- Mitral valve
- Adrenal glands
- Nasopharynx
- Spleen

- Taste buds
- White blood cells
- Endocrine system
- Lymph nodes
- Veins
- Epidermis
- Femur bone
- Olfactory nerve
- Uvula
- Retina

Those are the names of only 15 of those components, and there are over 1,000 pages of an anatomy book's body parts still to go!

Included in those pages are descriptions of the 79 different organs within the human body, all of which are basically interdependent. Every organ relies on the vital organs to function. For example, the lungs need blood. Blood needs the heart. The heart needs the liver. The liver needs the kidneys, and so on throughout the entire body. Vital organs are exactly that – vital. This means they could not have appeared one at a time. They must function together in a complete system, or the organism will die.

There are many organisms that do not have the same body parts as humans. For example, a fish has gills instead of lungs. But does that mean the DNA for gills can turn into the DNA for lungs? A snake has no arms or legs. But does that mean the DNA for nothing can turn into the DNA for arms and legs? Ridiculous!

If an organism got along fine without a liver, and the DNA for a liver mutated into existence in one of its offspring, what would the new liver possibly provide that was not necessary before? How could blood vessels, nerves and duct lines all perfectly connect to and from that newly mutated liver?

How about the kidneys? Could kidneys come from one massive mutation? A single mutation would require blood vessels, nerves, and kidney ducts to suddenly come into existence all at the same time, and immediately begin functioning perfectly. But that could only happen if the necessary DNA also happened to suddenly appear. Is that possible? Definitely not!

But could kidneys occur gradually over a long period of time? The answer is no. Non-working, partial organs getting passed down from generation to generation would be a huge burden. They would not aid in the survival of the organism in any way. Even if it were possible (and it isn't), wouldn't there be all sorts of partial organs existing in every living creature today? Yes, definitely. But surprisingly, there are basically zero.

Consider the reproductive organs. Could the DNA for male organs somehow appear accidentally in some sort of "pre-animal" that did not previously have that genetic code? How about the DNA for female organs? Where would that have come from? Could some sort of being that had no sex organs at all somehow mutate sex genes into existence and then start producing males or females – when there had been no such thing as males and females previously? The answer is simple. There is zero chance of such a thing ever happening.

This argument can continue using every vital organ in every living organism. How did prior generations survive without organs if future generations could not survive without them? Why do organisms living today not have previously unknown, partial organs mutating anywhere in their bodies?

The DNA for organs and other body parts could not have come about from random gene scrambling. The complexities involved eliminate any and all possibilities of their having appeared accidentally, whether one by one or all at the same time.

Chapter 11

THE FIRST HEART

Eleven

THE FIRST HEART

We cannot teach people anything; we can only help them discover it within themselves.

– Galileo Galilei

The human heart is a pump that moves blood through the arteries and veins. This incredible organ starts beating in the womb and continues beating nonstop for a lifetime. The heart typically beats about 36 million times per year, while blood pressure and pulse rate are maintained at required levels. All of this can only happen due to the existence of the complicated programming contained in DNA.

The heart is also comprised of two halves – the left and the right. Each half has an upper part called the atrium and a lower part called the ventricle. Blood enters the right atrium, is pumped to the right ventricle, and then to the lungs where it is re-oxygenated. The blood then returns to the heart via the left atrium, is pumped into the left ventricle, and then out through the aorta to the rest of the body. Incredible, isn't it?

Yet, the evolution story claims this highly involved, intricate mechanism came about by accident. Obviously, an entire heart could not have happened all at once, so the story claims it must have started out as something else. But what could possibly mutate itself into a heart? The answer of course is nothing, as we will soon see.

First of all, the heart needs internal programming to regulate beats per minute and blood pressure. This requires the use of sensors, monitors, and signaling devices via a system of nerves tied directly to the heart from the brain. The evolution story suggests that this complicated programming came about all on its own, which is utter nonsense. Such systems are extremely complex and could never have happened accidentally.

The evolution story would also require many other body parts to already be in place, just waiting for a heart to appear. But how could arteries, veins, capillaries, etc., have functioned without a heart anywhere to be found? What would they have done before the heart appeared? Furthermore, the heart would need a steady supply of oxygenated blood. So, which came first, the heart or the blood?

As we saw in Chapter Seven, even mutations occurring several times per second for *billions and billions* of years would not result in anything even remotely capable of producing life. Yet, we're supposed to believe that random mutations could result in a complete, fully functioning, perfect heart in just the right place, and with the necessary blood, arteries, veins, etc., already in place? Ridiculous!

The evolution story wants to have it both ways, claiming that a massive number of mutations occurred continually throughout the past – and that they mainly created benefits. The story not only has to ignore scientific probabilities, but it must also admit that mutations are very rare – *and* that they overwhelmingly result in defects. There are no new biological systems currently in preproduction stages in any living beings anywhere. None. How could that possibly be?

Despite evolutionists' claims, probabilities dictate that random mutations could not even begin to produce as little as a single organ in any life-form that has ever existed. Every individual body part has zero possibility of having occurred accidentally. This is especially true for the heart with its multiple chambers, muscles, valves, nerves, and perfectly placed blood vessels, among many other things.

Chapter 12

THE FIRST BRAIN

Twelve

THE FIRST BRAIN

I think, therefore I am.

– Rene Descartes

The evolution story claims that a single cell formed on its own and then accidentally came to life. It then supposedly proceeded to turn into some sort of brainless creature, which then mutated its DNA to form the very first brain cell ever. That definitely sounds like a harebrained concept, but we can analyze it anyway, just as a mental exercise.

To begin with, since one brain cell would be capable of very little thought, mutations would need to repeatedly occur until a tiny brain of some kind was in place. But in order for that to happen, those mutated brain cells would have to *accidentally* organize into the equivalent of computer hardware and software! First, the DNA would have to be arranged perfectly, then the brain would need to form into hardware, and the programs the brain contained would need to form into software. But the odds of brain DNA appearing accidentally are zero.

Ultimately, the senses of sight, smell, taste, touch, and sound all require communication lines that are wired directly to the brain. Could this complex circuitry accidentally appear at each of these sense organs? Could the ends of the wires accidentally terminate at the exact locations in the brain where the hardware and software accidentally happened to be? Could all of this simply be due to pure luck? Not a chance! The necessary DNA would have to be preprogrammed and perfectly arranged ahead of time or it could never happen.

The brain's biological equipment is so incredibly complicated that it defies a simple, straightforward description. For example, the largest part of the human brain is called the cerebrum, which is made of two hemispheres – the left and the right. It also contains four lobes – the frontal, temporal, parietal, and occipital, with each being split between the left and right hemispheres.

Half of the frontal lobe is in the left hemisphere, and half is in the right hemisphere. The same applies to the other three lobes, making the two hemispheres basically mirror images of each other. The two hemispheres are connected by large bundles of nerves (very much like highly intricate electrical wiring) allowing instantaneous communication between the various parts of the brain. Interestingly, despite their outward similarities, the right hemisphere is the emotional, artistic half of the brain, and the left hemisphere is the logical, reasoning half.

Even more interesting, each of the four cerebral lobes has multiple assignments. For example, one of the duties of the occipital lobe is analyzing the sense of sight. Light enters the eye, causing electrical impulses to travel along the optic nerve to the brain. Incredibly, the occipital lobe is programmed to receive and process shapes, motions, and colors. Together with other areas of the brain, the occipital lobe is capable of recognizing food, predators, prey, the opposite sex, and everything else that is seen by the eye.

As a matter of fact, you are reading these words right now thanks to the light leaving this page and entering your eyes, which in turn stimulate nerve cell transmissions to your brain.

Those nerve cell transmissions are then interpreted by your occipital lobe and other areas of your brain.

Even if a creature has eyes, if it doesn't also have sight-recognition programs in its brain, it most likely would not survive very long. For example, it would not know to be afraid of predators and could quickly get eaten. In addition, certain animals would not be aroused by the sight of the opposite sex for mating and would not propagate the species. They also would not know what to eat and could eventually starve to death.

It should be obvious that none of the genetic code for such hardware or software could have randomly mutated into existence. The DNA required is far too long and much too complicated. As our earlier probabilities discussion revealed, the tremendous complexity of the brain could not have resulted from genetic code shuffling – not a chance. Genetic code does not occur randomly, no matter what the evolution story might claim.

The brain receives and interprets signals of light, sound, taste, touch, and smell, and has astounding memorization and logical thought capabilities. Ultra-complex DNA creates ultra-complex brains that are intentionally and perfectly designed into every living creature on the planet. There is zero possibility of that happening through any process that involves zero brainpower.

Chapter 13

THE IMMUNE SYSTEM

Thirteen

THE IMMUNE SYSTEM

The best defense is a good offense.

– Jack Dempsey, boxing champion

Inside every single one of us, an immune system is fighting off bacteria, viruses, and cells that have become infected or have turned cancerous. If not for our immune systems, any and all invading microorganisms and cancer cells would be able to rapidly multiply and overwhelm our physical bodies, quickly killing us. Fortunately, we have a phenomenal seek-and-destroy defense system working within us to keep us alive and well.

The evolution story's version of how such a system came about raises several questions. To begin with, how could the earliest animals have survived microbial attacks if immune systems did not yet exist? How could they have survived later on when their immune systems were supposedly still undergoing development and were not yet anywhere near functional? Wouldn't they all have been killed by invading microorganisms such as viruses, causing the species to quickly go extinct?

The answers lie in the design of DNA and the complexity of the immune system itself. For example, here is a short list of only a very few of the components involved in the immune system:

- Antibody
- Cell-mediated response
- T cell, helper
- Suppressor cell

- Flaviviridae
- Opsonization
- Unique antigen
- Response, primary
- T cell, killer
- Eosinophil
- Neural regulation

In addition to the above, lymphocytes play a major role in the immune system. They are white blood cells and can come in the form of B cells, which produce antibodies that attack foreign invaders such as viruses. Lymphocytes also come in the form of T cells, which attack foreign cells, and also attack our body's own cells that have become infected or have turned cancerous.

Another very important component of the immune system is the macrophage. It has multiple assignments, but is most critical in the immune system's response mechanism. When a macrophage encounters a foreign cell, such as a bacteria cell, it proceeds to engulf it, then breaks it down into its various parts. The macrophage then identifies the bacterium's genetic markers, extracts them, and places them on the outside of the macrophage's cell wall. This is quite the undertaking. But wait, there's more!

After extracting a bacteria cell's genetic markers, the macrophage sends out a chemical message that activates the helper T cells. The helper T cells then send out messages to activate the B cells. The B cells then spring into action and travel to the macrophage to read the bacteria's genetic markers adhered to its surface. The B cells then produce antibodies that are programmed to find and latch onto bacteria with those exact genetic markers. An antibody does this by trying its "key" in a bacteria cell's "lock."

If there is a match, the antibody latches onto it, then sends out a signal to the macrophage to seek and destroy that particular bacteria cell! This is all highly organized and coordinated, much like a military unit with specialized detachments, each with specific instructions to carry out.

Amazingly, this is just one method the immune system has of destroying bacteria. There are many other mechanisms within the immune system that are constantly working to protect the body. Such intricate and complex processes could never happen via random mutations. Every component involved would serve no purpose until the entire overall genetic code was fully designed and up and running in absolute perfection. As shown in earlier chapters, the odds of any system's DNA developing accidentally are zero, and that includes DNA for the immune system.

Functional design had to have been intentionally and meticulously encoded into immune system DNA. The numerous biological components involved could not have come about accidentally, nor could they work in concert accidentally. Randomness has no chance at all of creating something so amazingly complicated, so perfectly tactical, and so incredibly perfect.

Chapter 14

THE SENSES

Fourteen

THE SENSES

*Design is not just what it looks like
and feels like. Design is how it works.*
– Steve Jobs

The Sense of Sight

According to the evolution story, the DNA necessary for light recognition began appearing one small segment at a time. Then a cell somehow mutated the genetic code into place for a primitive form of light recognition, such as for a plant bending toward sunlight. Then future generations allegedly scrambled their DNA on a regular basis, turned into animals, and somehow developed two fully functioning eyes that could focus, rotate in their sockets, adjust to brightness and darkness, etc. This is nonsense, of course. But let's look into it and see what we find.

Let's first imagine the DNA for a single light-sensing cell defying all odds by mutating into existence in a previously sightless creature. One cell alone, however, would not make it possible for the organism to see. The DNA for a huge number of additional cells would need to mutate into existence, creating several mechanisms that then would have needed to come together to form a highly complex overall system.

In advanced organisms, light gets partially focused as it goes through the cornea, then gets even more focused as it travels through the lens. Then photoreceptors in the retina send electrical signals to the brain along the optic nerve. Once these signals arrive at the brain, they are analyzed and processed by image-recognition programming.

The biological wiring from the eyes to the brain, and the brain's light-interpretation programs, would all have to exist in the same location and at the same time before a sense of sight could provide any benefit. But the odds against the accidental, timely occurrence of error-free DNA would make the haphazard manufacture of such a complex system impossible.

The first creature with vision would also somehow have to know exactly what to do when it suddenly could see things in its environment. It would need programs allowing it to recognize, seek, and remember certain shapes as food, as well as recognize and be attracted to the opposite sex. It would also need programming to recognize and fear predators and know how to evade them, and many more such programs.

It should be obvious at this point that the extremely long, highly complex DNA necessary for a functional sense of sight could not have come about by blind luck!

The Sense of Hearing

The outer ear comes in a set of two, and is not very complex in appearance. However, it is interesting that each outer ear just happens to be situated over a hole in the skull. If this had been caused by mutations, it would have been quite fortuitous, since truly random mutations would just as likely have grown ears on shoulders, ankles, or wrists, with one ear here and five ears there.

Not only are there two outer ears, but there are also two inner ears which are extremely complex, with mechanisms that work together to transform sound into electrical signals that are sent to the brain. Programs in the brain then analyze the signals appropriately. Listed below are just a few of the components that work together to make hearing possible:

- Membrane, tectorial
- Anvil
- Round window
- Keloids
- Fossa triangularis
- Oval window
- Upper channel
- Receptor cells
- Nerve, vestibular
- Incus
- Nerve, cochlear
- Eustachian tube

According to the evolution story, there was a time when none of the above items existed. Then somehow one day the DNA that creates them suddenly mutated into existence. This would mean that the DNA for the areas of the brain that interpret sound also happened to accidentally mutate into existence. However, since even a small amount of genetic code cannot occur on its own, as shown in earlier chapters, the vast amount of DNA required for fully functioning ears absolutely could not happen randomly.

Some creatures might have less-complicated vibration-detection systems, which the evolution story claims evolved into today's highly complicated, fully formed hearing systems. However, "less complicated" does not mean "capable of mutating into something massively more complex and entirely different." For example, vibration-detecting DNA as found in snails could never turn into the highly complex, double-ear hearing DNA that exists in mammals. There is simply no comparison and no logical pathway for such a thing to occur.

The Sense of Touch

An electrical power company has generating plants, transmission lines, switches, and monitoring devices, all designed to efficiently deliver electricity to homes, businesses, and factories. This takes place through a highly complex system of equipment interconnected through complicated wiring networks.

Remarkably, the human body has nerve lines laid out to form a complex electrical system complete with generators, transmission lines, switches, and monitoring devices, all designed to efficiently deliver electrical signals throughout the body. This takes place through a highly complex system of equipment interconnected through complicated wiring networks. In addition, the central processing unit - the brain - has receivers, transmitters, and signal-decoding hardware and software that connect to every part of the body.

In biology textbooks, nerve cells are described as being incredible, ingenious, and ultra-complex, which are all excellent descriptions! Nerve cells create their own electricity, transmit messages to the brain, and receive messages back from the brain. For instance, if someone accidentally touches a hot stove, the receptor nerves in the fingertip will instantly send signals to the brain. The brain immediately interprets those signals and sends messages to the hand, wrist, and arm to reflexively jerk the hand away – all in the blink of an eye!

How could such a complex sensory system ever have come into being? Could DNA accidentally appear one mutation at a time and create the complete code for the spinal column, which provides a protective conduit for the nervous system's main transmission line? Could all of the other nerve lines running throughout the body also occur by random chance? Absolutely not! The spinal cord runs through the spine, which is an ideal skeletal support structure. From there, the nerve lines spread out from the spinal cord to every part of the body.

The DNA that creates nerves could not accidentally mutate into existence, producing millions upon millions of nerves in the precise locations necessary, let alone with the correct wiring to the sections of the brain that control those areas.

No matter where a spider might be crawling on your skin, you can definitely feel it. This is because sensory nerves in the skin are linked to the body's neural network system, which is connected to the brain. The brain is designed with built-in programs that interpret various sensations, such as those caused by insects crawling on the skin, and even include the exact location of the offenders.

Miles and miles of nerve lines run throughout our human anatomy exactly where they are needed. If that had come about from random mutations, wouldn't there also be many miles of random, *unnecessary* nerve lines running from nothing, to nothing? Instead, miles of the right nerves and only the right nerves exist, to and from the exact locations intended.

Strange, isn't it, how absolutely perfect everything is? It might even give you goosebumps if you really think about it.

Chapter 15

THE FIRST BONE

Fifteen

THE FIRST BONE

Men occasionally stumble over the truth
but most of them pick themselves up and
hurry off as if nothing happened.
– Winston Churchill

Bones are much more complicated than might be expected. Just as with any mechanical structure, bones oftentimes have to operate under strenuous conditions. These stresses may include heavy lifting, while simultaneously twisting and turning. Bones need substantial strength characteristics designed into them.

Bones are made of living components that all work together. Our bones contain an outer membrane with blood vessels and nerves. This outer membrane covers hard bone, which may house either bone marrow or spongy bone. What an amazing design!

An adult human being typically has 206 bones altogether. There are a wide variety of different bone types, and they all have specific shapes and lengths. Could it possibly have been random events that produced 206 separate bones that fit together exactly, as though by design? Could "accidental" bones attach to each other and to all the various muscles simply by chance?

Consider that our upper and lower jaw bones firmly anchor each of our 32 teeth. When viewing x-rays, we see that every tooth has its roots sunk far down into the jaw bone or high up into the skull's maxilla bone. Incredibly, every one of the upper 16 teeth partners nicely with the 16 equal but opposite lower teeth. Our teeth also contain enamel, which happens to be the hardest substance in the entire body. Could any of that have been the result of random gene scrambling? How is it that the genetic code for a complete set of teeth even exists?

How about bone marrow? Amazingly, our bone marrow produces approximately 500 billion blood cells per day. It also plays an important role in our immune system. How could something as complicated as bone marrow just happen to reside inside our bones? As shown in earlier chapters, it could not have come from the accidental appearance of randomly occurring DNA.

If life had come about the way the evolution story suggests, the very first bone would have been quite a curiosity. Imagine some sort of boneless living thing, such as a jellyfish, suddenly producing an offspring that had a bone. Since it was the first bone ever, it would have served no purpose. It would have had no tendons or muscles attached to it since they didn't exist yet. It would have done nothing and supported nothing.

Could that first-ever, useless, dead-weight bone have been passed along from one generation to the next, waiting for a second bone to pop into existence? No. A second bone would result in twice as much dead weight, with absolutely zero survival advantage – especially with no connective tissue to hold the bones together. Quite a thick-skulled concept!

But let's continue our analysis of existing bones and their overall skeletal structure. To begin with, consider that a single human foot has 26 separate bones interconnected by 33 joints and over 100 ligaments. There are also a number of muscles and tendons involved. This incredible design also includes nerves and sensors which connect to the brain, allowing it to maintain the body's upright position, even when running at a full sprint!

With that in mind, let's imagine what the first animal with a foot might have been. Suppose it was a fish, since that's what the evolution story's caricatures show – a fish crawling out of the sea on little fin-feet. In some parts of the world, there are indeed catfish that can travel short distances on land by squirming along using their fins. But does that mean those squirmers evolved via mutations? Absolutely not! The ability to squirm along on land requires very specific genetic rearrangements not possible via random DNA scrambling.

A fin could never turn into a foot. There are no DNA commonalities, and the genetic changes required are too vast to allow any such thing to occur accidentally. Furthermore, a fish tail would have exactly the same odds of randomly becoming a foot, and you never see a foot-tailed fish hopping upright along the beach, do you?

A catfish is not going to turn into anything other than a catfish. It might be a small or a large catfish. It might be a light or a dark catfish. It might be water-bound, or it might be a water-and-land catfish. But they are all catfish, and they will never turn into anything other than catfish. Even the land-rover version is still a catfish and has fins just as all catfish do. It is simply able to wriggle to a new pond if its old pond dries up.

Instead of a foot bone, could the evolution story's very first bone have been a rib? Definitely not. There would be zero chance of the DNA for a rib occurring on its own. What chance would the DNA for a single rib have of accidentally changing into the DNA for a ribcage? Better yet, what are the odds of the DNA for a perfectly formed, complete skeleton accidentally mutating into existence by randomly producing one bone at a time? At what point did the DNA for bone marrow mutate into existence? All this "evolving" doesn't have a leg to stand on. In the end, it's simply a boneheaded story, and nothing more.

There is zero chance that the skeletal system happened accidentally. There would need to be an absolutely massive expansion of the required genetic code, including the DNA for all the necessary muscles, tendons, and nerves. There would also have to be new areas added to the brain to control the muscles in the arms, legs, torso, etc. Could that happen randomly? No way. There is no possibility that even one part of the skeletal system happened by accident, let alone the associated support systems.

Up next, we'll investigate "old" bones. Is it possible to determine dates and reconstruct history by examining bones found buried around the world? The answers may surprise you!

PART III

Here and Now

Chapter 16

THE FOSSIL RECORD

Sixteen

THE FOSSIL RECORD

You can fool all the people some of the time, and some of the people all the time, but you can't fool all the people all the time.
– Abraham Lincoln

No examination of the evolution story would be complete without an analysis of the fossil record. Long heralded as the story's crown jewel, evolutionists fervently believe that fossils provide definitive proof of evolution.

However, there is a very important fact that evolution scientists typically fail to mention – and most people have never even heard about. Quite surprisingly, small fossils can form in as little as three years! It is also not very well advertised that most fossils can form in fewer than 100 years regardless of size. That's right, fossils might not be anywhere near as old as commonly assumed.

The most common form of fossilization comes about from a process called permineralization. This occurs when a buried bone's many inherent air spaces are slowly but surely filled in by minerals in moist soil, turning the bone into a rock-like substance. This process, under the right conditions, can occur fairly quickly – geologically speaking.

Of course, this goes against the general perception that it takes millions of years for a fossil to form. To the contrary, fossilized bones that are found today might very well be from animals that died as recently as the 1900's. Of course, once fossilized, a specimen could conceivably last millions of years, but that does not at all mean that every fossilized bone is a leftover from prehistoric times.

Here is another very interesting fact: A fossil's actual age has no bearing on whether it is related to other fossils that existed earlier or later in time. In other words, apparent order of appearance does not in any way prove descent by family tree. If there is a lower layer of sediment that contains fern fossils, and the next layer up contains sea creatures, does that mean that the ferns turned into jellyfish? Could it prove that the jellyfish turned into human beings? Not a chance in the world!

As a simplified example, if a rabbit died and was buried in the year 1600, then a cat died and was buried above the rabbit in the year 1800, and their remains were dug up in the year 2000, would that prove that a rabbit had turned into a cat? Of course not. Even if a rabbit died 2 billion years ago, and a billion years after that a cat died, the odds remain essentially zero of any evolutionary connection whatsoever. The probabilities of such random genetic changes actually happening are infinitesimal.

Regardless, evolutionists still use fossils in an attempt to prove that one species can give rise to a completely different species. Naturally, they want to start their investigation only after life has populated the entire planet with continents full of plants and animals. So, let's investigate their theory.

We'll begin by discussing fossilized horse bones. In the 1800's, scientists unearthed some skeletal remains of small horses. They later estimated those horse fossils to be 50 million years old. They further assumed that the small horses had turned into modern-day, larger horses.

However, the evolution story is saddled with explaining why the little horses didn't turn into something more like Godzilla, instead of just becoming bigger horses. After all, 50 million years is an extremely long time, especially when they claim that a so-called missing link creature not only turned into apes, but also into humans – *and* that it happened in only the last 5 million years! So, it took 50 million years for a small horse to turn into a big horse, but only 5 million years for a missing link creature to turn into apes and humans? We'll let you do the math.

While we're on the subject, let's investigate the evolutionists' contention that humans and other primates have a fossil history that supposedly proves they share a common ancestor. While humans and apes do have some similar DNA, there are also significant differences. For starters, apes have 48 chromosomes, while humans have 46. Would it be possible for four "pre-ape" chromosomes to combine and form two human chromosomes? That would be an extremely lucky mutation combination for all of us humans. Or was it vice versa, with "pre-human" DNA mutating into ape DNA? Either way, neither luck nor mutations had anything to do with it, as we're about to see.

Current estimates state that human beings and apes have approximately 95% common DNA. However, how surprising is it that creatures with a heart, lungs, skeleton, eyes, ears, internal organs, blood, veins, immune systems, etc., have much of the same DNA that creates those items? This raises a question: why didn't the 95% common DNA also change if mutations are indeed random? How could the 95% have remained unchanged – especially over the course of an alleged 5 million years?

Furthermore, 95% common DNA indicates that 5% of the DNA is not the same. Since human DNA has 3 billion amines, a seemingly minor 5% actually equates to a huge amount of DNA that is *not* common – a very significant 150 million amines! As shown in Chapter Seven, even extremely miniscule beneficial changes in DNA via accidental means are impossible. Therefore, such a gigantic evolutionary leap could not have happened accidentally under any circumstances.

Years ago, many evolution scientists grew tired of trying to provide rational explanations for the massive gaps involved in the evolution story's sudden appearance of new DNA. Since there is scant evidence of a slow, steady transition from one species to an entirely different species, they proposed that whenever a new species occurred, it must have been caused by something they called "punctuated equilibrium," which was supposedly the sudden appearance of large amounts of new DNA. One day it was DNA for the missing link, and the next day it was DNA for human beings. Presto change-o, just like magic. Most scientists don't believe in magic, but it seems evolution scientists do!

These same scientists also claim punctuated equilibrium didn't happen very often, and that's why no one has ever witnessed it occurring. In reality, the punctuated equilibrium story is totally lacking in logical analysis. There is no chance that the DNA for any single piece of anatomy happened accidentally. There is also no chance that any species randomly changed into a totally different species one amine at a time, and certainly not via one big old mega-mutation.

Furthermore, if the time it took for the entirety of all punctuated equilibrium to occur since day one added up to just a few days in all, it is especially impossible to have been accidental.

DNA cannot occur on its own, even in very small amounts, which means that no form of life could possibly have come into existence on its own. Genetic variations that create entirely new species have no chance of happening randomly, whether by gradual change, punctuated equilibrium, or whatever the primordial soup of the day might happen to be.

Chapter 17

CARBON-14 DATING

Seventeen

CARBON-14 DATING

The best part of our knowledge is that part
which teaches us where knowledge
leaves off and ignorance begins.
– Oliver Wendell Holmes

As we learned in the previous chapter, many scientists believe that a fossil's alleged age has significant bearing on the evolution story's validity. So in this chapter and the next, we will investigate the electronic equipment used by scientists in their attempts to assign dates to bones found buried around the world.

We will set the stage by discussing the Earth's atmosphere, most of which is within 12 miles of the planet's surface. The air we breathe is comprised of approximately 78% nitrogen, 21% oxygen, 1% argon, and very small amounts of other gases, such as carbon dioxide at 0.04%.

Nitrogen atoms have a well-balanced seven protons and seven neutrons – normally. But an important point to know about the Earth's atmosphere is that it regularly gets bombarded by cosmic rays from outer space. And those rays occasionally cause a nitrogen atom's proton to turn into a neutron. This altered-state nitrogen goes up a notch from seven neutrons to eight neutrons, while simultaneously going down a notch from seven protons to just six protons.

Well, in the world of science, it's the protons that count, no matter how many neutrons there might be. This means that any atom with six protons is called *carbon*, regardless of how many neutrons it has. So when a nitrogen atom loses one of its seven protons and ends up with just six protons, it is then called *carbon*, even though it is an extremely rare form of carbon that has a surprising eight neutrons instead of *carbon's* usual six neutrons.

The former seven-proton, seven-neutron *nitrogen* atom that becomes a six-proton, eight-neutron *carbon* atom is actually a radioactive isotope and is referred to as the ever-infamous carbon-14. But don't be intimidated by the term "carbon-14." It is called that because there are six protons and eight neutrons, and six plus eight equals 14. Not too terribly complicated.

After carbon-14 is created, it gets swirled around by atmospheric winds. Some of it eventually ends up combining with oxygen, resulting in carbon dioxide, which is a molecule with one carbon atom and two oxygen atoms. The oxygen stays normal, but the carbon-14 component remains mildly radioactive due to the two extra neutrons.

OK, now we'll see how this applies to plants and animals. Plants take in carbon dioxide through photosynthesis. Although the great majority of the carbon dioxide is in its normal form, a tiny amount is carbon-14. Therefore, the carbon-14 gets absorbed into the plants' tissues. The plants then get eaten by animals, and the animals then get eaten by other animals. So a sparse amount of carbon-14 is lightly present in all plants and animals.

However, as soon as carbon-14 is formed in the atmosphere, it starts to go through the very gradual process of radioactive decay, slowly turning back into nitrogen. The carbon-14 in plants and animals also slowly turns back into nitrogen. Eventually, over the course of about 60,000 years, carbon-14 completely turns back into nitrogen, with seven protons and seven neutrons, leaving no trace of its former radioactive-carbon-14 self behind.

And now, here is where the use of electronic equipment comes into play. Scientists use electronic meters to measure the level of carbon-14 and compare it to regular carbon in long-dead plants and animals. From this they believe they can determine a specimen's age going back tens of thousands of years. However, there are a number of difficulties with this concept.

First of all, only about one in every trillion carbon atoms is carbon-14. And since carbon dioxide is only 0.04% of the Earth's atmosphere, that means carbon-14's presence is a miniscule one atom out of every 2.5 quadrillion atoms in the air. Therefore, scientists' measuring devices can only detect carbon-14 in things that died a theoretical maximum of 30,000 years ago, and then they "lose the scent." This is because the level of carbon-14 at that point is already so close to zero that meters don't have the sensitivity to detect it.

Over the years, scientists have occasionally found fossils that nicely fill in gaps in museum displays. So they only measure for argon (next chapter), then say the fossil is millions of years old and call it good. However, later testing sometimes showed the presence of carbon-14 in fossils that scientists had estimated to be millions of years old! But carbon-14 is basically undetectable after 30,000 years. So how in the world could it be found in supposedly multi-million year old fossils?

Well, many scientists try to explain this away as being caused by laboratory contamination or quirks in their measuring devices. But there is much more involved than just general inaccuracies and equipment lapses.

For starters, the historical rate at which carbon-14 was created in the atmosphere cannot be precisely known. This is due to constant variations in the presence and strength of the cosmic rays entering the atmosphere at any given time. Also, Earth's magnetic field fluctuates significantly, which affects carbon-14 formation. Long-term absorption into the world's oceans further influences atmospheric carbon-14 levels. These factors contribute substantially to the uncertainty of historical carbon-14 levels.

All of this must be taken into consideration since the amount of carbon-14 in the air determines its concentration in plants and animals at any given time. We simply cannot know what the level of atmospheric carbon-14 was, due to the variables mentioned above. Machines might attempt to tell us the present-day amount of carbon-14 in a historical specimen, but scientists then have to make many unprovable assumptions to achieve nothing more than an age guesstimate.

It seems logical to assume that near the beginning of the Earth's existence there would have been zero carbon-14 in the air. Then, once there was nitrogen, the cosmic rays would have started producing trace amounts of carbon-14. It also seems logical to assume that the amount would vary over time as new carbon-14 was added to the existing carbon-14, which was constantly decaying back into nitrogen. Consequently, this would make it impossible to know what the concentration was at any given time in history, other than day one, when it would have been exactly zero. But science alone cannot tell us when day one was. Quite simply, there is no way to know the age of fossils or Earth's creation date through the use of machines.

Don't believe it yet? Please continue reading...

Chapter 18

"BILLION" YEAR OLD ROCKS

Eighteen

"BILLION" YEAR OLD ROCKS

Science is a way of thinking
much more than it is a body of knowledge.
– Carl Sagan

In the previous chapter, we learned that some fossils contain carbon-14, which is an isotope that can be detected by using electronic devices. But did you know carbon-14 is the *only* item in a fossil that can be measured to estimate its age? That bears repeating. In the fossil itself, electronic devices can only get a reading on carbon-14, and nothing else, when estimating how long ago a creature might have roamed the Earth.

We also learned that carbon-14 dating can only go back a maximum of 30,000 years. So how can scientists claim that those fossils are millions or even billions of years old if the carbon-14 they contain would have been undetectable after 30,000 years? Are you ready for the answer? It's because they don't use the fossils at all! Believe it or not, they take measurements of the rocks or sediment *around* fossils and then assign a supposed age.

Oddly enough, the particular element they typically look for in the rocks around a fossil is argon. As we know from the previous chapter, argon is the third most prevalent gas in the Earth's atmosphere. Evolution scientists propose that this large amount of argon originally seeped out of rocks and into the air. And since rocks contain potassium which very, very, very slowly turns into argon, they theorize that over the course of billions of years, the potassium-turned-argon supposedly escaped from the rocks and became the third most abundant gas in the atmosphere.

Contrarily, there are also scientists who say that the lattice structure of rocks *prevents* argon from exiting or entering the rocks. This would mean that argon is basically encapsulated in the rocks forever. Therefore, they theorize that because argon cannot enter or escape, it could be a good indicator of a rock's age. The more argon there is in a rock, the older they figure the rock is. For example, they postulate that if a rock originally had zero argon, and over a long period of time some of its potassium decayed and turned into argon, then the age of the rock could be determined.

Yes, you read that right! On the one hand, some scientists say that over long periods of time argon *escaped* from rocks, eventually creating the large amount of argon we now have in our atmosphere. Then on the other hand, they say argon is trapped inside rocks and *cannot escape*, and therefore could supposedly yield reliable age guesstimates. These same scientists also claim that the structure of rocks *prevents* argon from entering from the outside – even though argon is plentiful in the atmosphere and in seawater! Are you scratching your head yet?

It's another little-known secret that when the argon measurements don't come anywhere close to the desired results, some scientists simply ignore those readings and assign an age such as 75 million years anyway! This "best-fit" dating method might be used if the rock around a fossilized skull tested as being far younger than hoped for. At other times, they might assign an age that allows them to best fit a skull into a particular museum display. Besides, who would know the difference?

Also, as mentioned in the previous chapter, many so-called million-year-old fossils test positive for carbon-14. This is exasperating to evolution scientists, since carbon-14 measurements can only go back a maximum of 30,000 years, if that. So how in the world could it be found inside a fossil that is supposed to be *millions* of years old? Regardless, if any of that pesky carbon-14 does show up in their test results, they say those samples were affected by carbon-14 in the laboratory air. Then they speculate that their machines weren't recently calibrated, or perhaps the discrepancies were caused by operator error. Are you scratching the other side of your head yet?

How can anyone possibly know what happens to dirt and rocks over the course of thousands of years, let alone an alleged millions or billions of years? It is well known that floods have occurred all over the planet – including a potentially massive flood. And as you know, when a flood occurs, incredible amounts of soil get moved to new locations, thoroughly intermixing sediments from different eras in the process. So, isn't it possible that skeletal remains could end up in sediment that tests much older than the fossils' actual origins?

Speaking of soil and rock displacement, let's now consider the many volcanoes that have erupted over the entire planet for millennia. Oregon's Crater Lake was formed thousands of years ago when Mount Mazama erupted, pouring massive amounts of ash into the atmosphere. As recently as 1980, Washington State's Mount St. Helens erupted and spewed huge amounts of ash that circled the globe for years. These are just two volcanoes that emanated from one tiny portion of the planet, and both of them created significant soil strata changes in large geographic areas.

There have also been many relatively recent volcanic eruptions in the state of Hawaii. Some of these occurred only 200 years ago, yet the rocks they produced tested as being millions and even billions of years old! Yes, these age estimates were for 200-year-old rocks.

Because of this, many evolutionists agree their methodology might not be "rock solid." They will simply explain away irregularities by saying certain types of minerals within lava can yield erroneous readings. Of course, this only adds to their ongoing guesswork factor.

We've all seen news footage of volcanoes, cyclones, tsunamis, hurricanes and floods that have happened around the planet, all in very recent history. So how could evolutionists possibly know what all has happened since creation up to the present day? In actuality, they can claim whatever they want, but they are simply guessing – and in grand fashion. There are no samples of dirt or rocks that have been kept in sealed jars for millions or billions of years that can be used as test subjects. Consequently, there is no definitive way to accurately measure the age of dirt or rocks. It all amounts to speculation and little else.

In addition, there is no way of knowing how much argon was trapped in every rock on the planet when the Earth was created. There also is no way of knowing how much argon was in rocks that much later formed around fossils, or how much escaped from rocks, or how much seeped into rocks.

There is no way to determine via machines what happened in the distant past. Machines can only measure things as they are in the present – in the here and now – and even then are subject to calibration error, operator error, and laboratory error.

We have no technological way of determining when the Earth's exact starting point was, or what the geological conditions were at that time. Such facts and figures cannot be determined through the use of electronic devices. These mechanisms only allow evolutionists to make wide-ranging speculations as to what "might" have happened in the past, all based upon unprovable assumptions.

Bear in mind that the entire "age of fossils" debate is just a diversion from the scientific facts detailed earlier in this book. Life cannot create itself, and no species can change into a totally different species, even in a near-infinite number of years.

Note from the author:

Points of view sometimes don't make any sense, requiring logical gymnastics and mentally strenuous contortions. Even though evolution is the evolutionists' gospel, it definitely has exhibited far too many exceptionally erroneous suppositions, with explanations that are nothing but imaginative. The evolution story's weaknesses simply don't offer anything credible.

Chapter 19

VIRAL MARKERS

Nineteen

VIRAL MARKERS

In theory, there is no difference between theory and practice. In practice, there is.

– Yogi Berra

As discussed in Chapter Eight, a virus latches onto the outside of a cell wall, then injects its DNA into the cell's interior. The victimized cell then runs the viral DNA through the cell's ribosome and produces viruses, one after another.

However, there is a different kind of virus called an endogenous retrovirus (which is abbreviated ERV). When an ERV's DNA enters a host cell, it invades the cell's nucleus, goes to a preferred location in the cell's DNA, breaks it open, and splices its own genetic code directly into the cell's DNA. This is an absolutely remarkable undertaking!

It is also known that genetic code very similar to some ERV genetic code is present in both chimpanzee DNA and human DNA. What does this mean to evolution scientists? They postulate that an ERV must have inserted its genetic code into a "missing link" creature's DNA, and then that creature mutated one direction into chimpanzees, and another direction into human beings. The story goes on to claim that this is the reason chimps and humans both have these "accidental" markers. But as we learned in previous chapters, there are quite a few problems with the missing link story, including this one.

As detailed earlier, humans and apes do have a significant amount of common DNA. But does that indicate they evolved from a common ancestor, since there are some of the same "viral markers" in our genes? Absolutely not! A virus could not have been involved in any of this.

For starters, current estimates state that a missing link's entire body would have contained somewhere in the neighborhood of 37 trillion cells. But if the virus were to attack a kidney cell, or a thumb cell, or any cell at all among the entire 37 trillion cells available to choose from, the viral DNA would not get passed along to future generations. There is only one biological mechanism through which such a thing might theoretically take place: the ERVs would have to repeatedly attack the missing link's sperm cells or egg cells. But could that actually happen? No way.

There are about 250 million sperm cells released per sexual encounter, so it would be impossible for ERVs to make any headway at all against the sudden influx of such a massive, highly mobile, fast-moving army. And even if an ERV ever did latch onto a sperm cell, it would only have a one in 250 million chance of being the sperm cell that succeeds. So, that leaves egg cells as a possible target. But that is not how ERVs operate.

The ERVs could attack any of the missing link's 37 trillion cells. Probabilities therefore dictate that it would be a one in 37 trillion shot for an ERV to randomly happen upon an egg cell, let alone breach its formidable cell wall.

Consider that sperm cells experience first-hand what a difficult assignment it is to enter an egg cell – and that's with 250 million sperm cells all working on the same exact assignment at the same time! So what chance would a "lone ranger" ERV have at 37 trillion to one odds? Zero chance.

To have the same missing link ERV segments present in human DNA and chimp DNA, there would need to be many repeated, successful ERV attacks on that one particular missing link's family line, specifically targeting its egg cells. But that scenario is far beyond the realm of possibilities. Such a thing would not realistically happen even once, let alone over and over again with a wide variety of different ERVs, each of which would have bypassed 37 trillion cells in each of the "missing link's" ancestors. Additionally, there is no way a missing link could have escaped negative results from random, out-of-control ERVs attacking its entire body, with only one in 37 trillion ERVs ever finding their way to an almost impossible to penetrate egg cell.

OK, before continuing, we had best change gears for a moment and look at a computer programming analogy. Starting with a random sample, we'll use the sequence 01001010, which is binary for the letter "J." We'll also use the sequence 01000011, which is binary for the letter "C." Now, suppose we take the "J" and *randomly* insert it, say, two digits from the end of the "C." This results in 0100000100101011, which is "A+" in binary.

This is an important point: we started with "J," inserted it into "C," and ended up with something completely different, "A+." And no computer would be improved by randomly turning "JC" into "A+." For such a thing to be beneficial, it would have to be encoded into the programming ahead of time. This holds true for basically every code insertion imaginable.

Obviously, the random insertion of pieces of computer code directly into other programs would create havoc, severely hampering the program. However, a talented computer programmer would be able to determine where to insert lines of code and make all of the proper adjustments as necessary. But computer code, even if very minimal in length, cannot be inserted randomly into a program here and there and just happen to be beneficial. To the contrary, it would have to be well thought out and intentional, or it would be destructive. In a like manner, only the designer of genetic code would know where to locate each amine, how to precisely tweak the surrounding DNA, and how to manipulate the entire gene.

Also, understand that any insertions of ERV genetic code into missing link DNA would have to be passed along to future generations with only targeted changes in those ERV segments. Meanwhile, massive, *150-million* amine alterations in the missing link's overall DNA would have to take place, with one branch turning into chimps and the other branch turning into humans. The new genetic code, *if* accidental, would need to align in precise order, with 150 million newly arranged bits of code "mutating" perfectly, and everything else remaining unchanged.

However, we know from earlier chapters that accidental transitions are not possible. Even if they were possible, wouldn't that mean that the "viral" DNA would have randomly mutated into something entirely different over the course of millions of years?

And now it's time for something very interesting. There are said to be about 100,000 segments of retrovirus DNA in human beings – comprising a stunning 8% of our 3 billion amines! Could that have all come about from random viral attacks? Definitely not, as we've been learning throughout this book!

How can anyone claim that so many impossible occurrences actually happened by accident? The DNA in question fits into the overall genetic design, with only specific alterations included as necessary in each individual species. That cannot be an accident in any way, shape, or form.

Chapter 20

GENETIC ENGINEERING

Twenty

GENETIC ENGINEERING

A good engineer is a person who makes designs that work with as few original ideas as possible. There are no sissies in engineering.

– Freeman Dyson

As we learned earlier, genetic code is comprised of four molecules called amines. These amines are arranged in sequences which form codons. The codons are then used by ribosomes to produce amino acids. The amino acids then form proteins, and those proteins construct you and me. Could this all be an accident? Definitely not.

These four miraculous amines not only perform the aforementioned feats, but even more astoundingly, many amine sequences serve a dual purpose! They not only form proteins, they are also simultaneously involved in very complicated gene control and regulation. This is truly incredible! How could a single set of amines perform two functions at the same time? The complexity is beyond belief.

As an analogy, consider the first page of Chapter 15, which contains regular messages in the standard writing. However, if it *also* contains a message along most of its left margin, it would either be intentional – or an accidental miracle! Of course, we know it certainly isn't accidental. The exact same reasoning is true for DNA. Its multi-messaging amines cannot possibly have occurred randomly.

As if that weren't enough, there are also genetic regulators attached to the *outside* of the basic genetic code. These are called epigenetic markers, and can turn genes on and off in ways so complicated it is not yet even close to being fully understood.

With all this in mind, let's investigate genetic engineering. Today's genetic engineers are capable of tinkering with a few extremely small segments of DNA in order to give some plants insect resistance, and others increased tolerance to frost. But is there more to it than meets the eye?

Operating on a very basic level, scientists perform genetic engineering by borrowing a small segment of DNA from one species and inserting it into another species. However, they most certainly don't insert *random* bits of genetic code *just anywhere*, then stand back to see what happens. No, these scientists realize it would be a waste of time to insert randomly arranged amines into random locations in DNA. Instead, they use portions of genetic code that already exist in nature, and insert them into specific locations in DNA. They would never bother going through the immense effort of trying to create even a tiny segment of DNA via original design.

Controversy understandably arises from making genetic alterations while not knowing how the sub-programming and epigenetics might be affected. Accidentally creating adverse reactions becomes an ominous possibility, and some negative results do appear to be occurring in some genetically modified products.

For several decades now, there has been great excitement among scientists to quickly identify the purpose of any section of genetic code, no matter how small. Why such a sense of urgency? Because they can obtain patents for any identified DNA segments! And with *huge* amounts of money potentially at stake, the research has been intense.

But this raises a question. Since evolution scientists believe billions of amines can line up randomly and form life, why don't they simply determine what each individual amine arrangement would accomplish? It certainly seems as though it should be easy enough, *if* you believe the evolution story. If something happened by accident, then designing it on purpose should be a piece of cake! The problem is, no one can tell what combination of amines does what, other than those that already exist in nature.

Therefore, why don't scientists just forge ahead and patent every possible amine combination and be done with it? The answer, of course, is because DNA is so incredibly complex that it would be impossible. A simple 27-amine segment, for example, would pose more than *18 quadrillion* possible arrangement combinations! Yes, 18 quadrillion. And that would be for just 27 amines of the billions of DNA's amines needed to create life. As you can see, those are some inconceivable numbers! But let's keep investigating, just for fun.

If you're still not convinced that tiny 27-amine segments cannot occur accidentally, let's bump things up a bit and look at slightly larger 140-amine segments. An example of just *one* 140-amine segment might be:

CATACGCCCTCGCGGCCTCACGGCCGATCTCCCTATAGAAATACATGCATGG
ATACATGCAGAAAGAAAGAACACCCGATCGATCGTACGCCCTATCGGCCGAC
CTATCTCACGCCCTATAGAAATATATGGATACATCA.

Now, imagine *every* possible 140-amine combination. As it works out, there would be over 194,267,000,010,010,100,110, 111,101,101,000,011,011,100,010,000,000,110,001,001,101,010, 011,101,000,110,001 different amine arrangements possible, and each would look much the same as our example above. And yes, that would be about two hundred thousand trillion trillion trillion trillion trillion different possibilities - just for a very minimal 140 amines with about 3 billion perfectly aligned amines still to go. No wonder every genetic engineer has chosen to simply copy the original designer's work instead. What a time-saver!

No human being has ever created functional genetic code of any significance from scratch. The best anyone has been able to do is copy – yes, copy – the genetic code for something such as a mouse or a sheep. On a lesser scale, there are some genetic engineers who simply take small DNA segments that exist in nature and insert them into other existing segments of DNA. Even then, they think they are being quite clever for performing maneuvers that are in reality pre-elementary in comparison to designing *billions* of precisely ordered amines with multiple layers of programming – starting from nothing.

When computer programmers copy someone else's software, they do not receive any praise for having done so. Quite the contrary. Original programming of any kind requires a tremendous amount of intelligence and know-how. Copying does not.

In the future, will the fossil record prove that today's genetically engineered plants accidentally mutated into existence? No, of course not. Any such altered genetic code will undoubtedly look like the work of rank amateurs who had no knowledge of basic programming or sub-coding whatsoever.

The fact remains that genetically engineered plants are currently being modified using already-existing DNA segments. In addition, these changes are being made with absolutely no interworking genetic-design considerations taken into account. When inserted into new locations, small alterations that are hoped to be beneficial and harmless could instead negatively affect all future generations – and to an extent that is completely unforeseeable.

Chapter 21

CONTROL SYSTEMS

Twenty One

CONTROL SYSTEMS

One test result is worth more than
a thousand expert opinions.
– Wernher von Braun

In this book's Preface, the author's "revelation moment" came when he was re-reading his arguments that are now included in this chapter. It is intended for any hard-core holdouts who are still wavering at this late stage!

Electronic control systems are designed to keep mechanisms operating within certain limits. For example, an automobile's cruise control system keeps a vehicle traveling at a chosen speed. It is a very complicated system that uses advanced mathematics programmed into a mini-computer and electronic devices, all of which have been carefully designed, manufactured, and installed. This is not something that can happen all on its own, completely by accident.

When this book's author was a college student, an engineering professor shared how his uncle owned a very large farm. While plowing the fields, the uncle had grown tired of having to turn around and look back to make sure the plow blade wasn't too deep or too shallow in the soil. So, the professor decided he would try to design a control system that would automatically control the plow blade's depth.

Unfortunately, the professor's first design attempt failed. The plow dug too far down into the soil, then jumped up to the surface, bouncing all over the place. After a great deal of double-checking, the professor discovered an error in his calculations. So he made what he thought were the necessary corrections and gave it another try, but there was no improvement. After several more days of intense analysis and design corrections, his invention continued to malfunction. So the uncle unfortunately had to go back to using his own natural abilities, and continued turning around and looking back in order to control the plow blade.

The uncle's brain, eyes, hands, etc. (which the evolution story claims happened completely by accident) allowed him to use his mental and physical capabilities to set the plow depth. On the other hand, the nephew college professor, with his advanced education, encountered great difficulty in his attempts to *intentionally* design a system capable of controlling the plow's depth. But how could an "accidental" uncle be able to do naturally what an educated professor's precisely designed devices could not? Let's find out.

There are many control systems (typically referred to as "homeostasis") constantly at work in all living beings. Human body temperature is maintained around the clock at a fairly steady 98 to 99 degrees. Other control systems monitor and control blood sugar levels and mineral levels such as sodium, potassium, and calcium. Still others control blood pressure, hormone levels, and protein levels.

Also, when a virus invades the body and multiplies to a dangerous level, there is a control system that raises the body's temperature to assist in fighting off the virus, returning the temperature to normal only after the viral threat has been reduced.

These control systems all work together in a wondrously coordinated system. None of these sophisticated control systems could have happened randomly, since they could not have come about from random DNA rearrangements. The complete programming for living organisms is essential for life and absolutely cannot occur little by little. There would be no benefit whatsoever from partial programs. The DNA had to be one hundred percent complete to be of any value.

Volumes of college textbooks are filled with information on man-made electronic control systems and the mathematical perfection required to design them. Similarly, the extremely complicated science involved in *biological* control systems had to be designed – but by a source of much, much higher intelligence. Life could not exist without the absolutely perfect genetic code for control systems already in place. Such systems could never have arisen from randomly shuffling DNA.

Chapter 22

BIG BANGS

Twenty Two

BIG BANGS

I don't know what you could say about a day in which you've seen four beautiful sunsets.

– John Glenn

Early on, evolutionists ignored the fact that there must have been a time when the universe did not exist, and then billions of galaxies, each with billions of stars, somehow appeared. Since they had no explanation whatsoever for this, an "exploding dot" story was started to ridicule them, kidding that a tiny spot created everything that exists.

But, since they had no alternative explanation, evolutionists came to embrace the concept. They speculated that not only was there a dot – but it appeared all on its own and then for some unknown reason exploded. This mysterious little speck eventually came to be known as the highly complicated point in the fabric of the time-space continuum that was punctured, and through which our universe poured out into being. Or something like that. And of course they theorized that the speck – and our universe – happened entirely by accident.

Ongoing speculation by other scientists argued that chunks of matter might have blasted into space from the exploding dot. This matter swirled around to somehow form billions and billions of stars, with each star being a gigantic nuclear reactor much like our sun. Then, just as magically, other celestial bodies somehow avoided the nuclear option and simply became cool planets that orbited those infernos. Then the billions and billions of stars and planets magically came together to form billions upon billions of separate galaxies. Just because.

As it turned out, the dot-exploding-into-chunks version of the evolution story fell victim to serious questions regarding the non-logic of its physics and astrophysics, among other things. It slowly faded away, but only after "new and improved" versions came along to take its place.

There were many big bang stories that followed, such as the possibility that our universe came from space dust that somehow exploded into existence. This dust supposedly swirled together to form stars, planets, moons, comets, asteroids, etc. Another version speculated that if the world happened accidentally, perhaps it was from some kind of primary explosion followed by an untold number of secondary explosions which formed all of the elements that exist in the universe.

Yet another version says there was no explosion at all. Instead, the dot simply expanded slowly but steadily, eventually forming our entire universe. Still another story describes the world as some sort of hologram – a massive illusion that came about accidentally via unknown means.

More recent versions claim energy was expelled by a big bang. This energy then floated together to somehow form protons, neutrons, and electrons, which then amazingly drifted together to form atoms, over the course of a very long time. Then, those atoms supposedly floated together and accidentally formed stars, as well as planets to magically orbit those stars.

There is other speculation that before a big bang could take place, a massive amount of subatomic components would have to come into existence all on their own. Those components would then need to migrate to a single location, arrange themselves in a favorable pre-explosion arrangement, and then "BOOM!" An unimaginably massive universe would somehow result, and life would be created from accidentally occurring molecules. As you might have guessed, there are many issues with this story also.

There are many contradictory, creative ideas available, postulated by astrophysicists, mathematicians, physicists, and other educated individuals. Some of them quarrel about how big bangs might occur, while others argue about whether big bangs can even happen at all.

Take a moment to consider our sun, which is a massive ball of energy. Now look at our moon, and all of the stars in the night sky too numerous to count. Is it possible that the existence of these celestial bodies is proof that they can be traced back to an exploding freckle in space? Definitely not!

In any solar system, in order for planets to be captured by their sun's gravity and settle into orbit, their speed, distance, and angle of approach would have to be ideal for their individual mass. Otherwise, each planet would either get pulled into its sun's nuclear furnace or go hurtling off into space. It's a very delicate balancing act that could not happen by chance.

The odds of a comet colliding with planet Earth in any given year are quite small. But it has happened on occasion over millennia, and could conceivably happen again someday. However, what are the odds of a moon-sized celestial body blazing by our planet, barely missing us, then slowing down and settling into orbit as Earth's second moon? Could that happen? The odds of any celestial body creating itself and then *accidentally* settling into orbit around Earth – or any other planet – are virtually zero.

Speaking of celestial bodies, a popular story about the origin of our moon speculates that billions of years ago the planet Mars was bombarded by comets and asteroids. This barrage eventually somehow broke off a huge chunk of the red planet. The chunk then supposedly shot out into the vastness of space and somehow accidentally hit the sweet spot and settled into orbit around the tiny speck called Earth – a "mere" 34 million miles or so away. That story is quite laughable, since the probabilities of randomly executing an exact-speed, perfect bullseye on a moving target from such a ridiculously enormous distance are infinitesimal.

So, to sound just a little less ridiculous, the story was changed to say that it was planet *Earth* instead of Mars that was bombarded by comets and asteroids, causing a huge chunk to eventually break off and become our moon. However, for orbital separation to occur, a massive blast would have to expend exactly the right amount of energy to propel that huge chunk away from Earth's gravitational pull to approximately 250,000 miles away and then "put the brakes on," or it would continue on out into space. Need we suggest that this entire *accidental* scenario is simply not possible?

As far as big bangs go, every theory out there has unresolved, nagging issues, such as the flatness of the universe, which is a highly unlikely result from an explosion. Another issue is that various regions of space do not display a common causation in terms of the speed of light and space-related physics.

There are many more big bang quandaries. For example, how many big bangs have occurred outside our realm of observation, beyond the reach of our telescopes, or maybe in other dimensions? Do big bangs pre-assemble themselves like Fourth of July fireworks displays, but with no intelligent input at all? For that matter, why couldn't a big bang explode from your belly button, or from between your toes, or from inside your nose?

Additionally, why don't scientists just go out into a field somewhere and create a mini-big bang? If big bangs are so simple that they can happen accidentally, why can't scientists easily create one from scratch?

How could a theorized pre-big bang's unimaginably massive amount of energy simply gather together and then randomly explode? Furthermore, why would energy even exist in the first place? Where could it have come from? How could energy explode and produce particles in the exact amounts and in the exact locations needed to then "accidentally" form atoms? Furthermore, how could those atoms end up in the exact proportions and exact locations necessary to "accidentally" form stars and planets by the billions which then formed galaxies by the billions? And lastly, how could a series of total accidents continue nonstop for billions and billions of years, to where you were created and are now reading this?

The odds of randomly hitting upon perfection in an ever-expanding universe, one galaxy after another, one star after another, one planet after another, and one moon after another – are zero. It is well known that there are billions and billions of galaxies, and each galaxy has billions and billions of stars. What might *not* be so well known is the fact that those billions and billions of galaxies could not have simply fallen into place by accident.

The probabilities against galactic randomness, interstellar randomness, solar system randomness, planetary randomness, eco-system randomness, biological randomness, cellular randomness, and genetic code randomness are beyond astronomical.

CONCLUSION

CONCLUSION

History shows that new ideas in science
most often come from brash youngsters,
mavericks, or rank outsiders.

– Fortune magazine, 1974

In order to properly analyze the evolution story, it is necessary to understand not only biology, but also related sciences such as the following:

- Programming fundamentals, which apply directly to the functions of DNA
- Probabilities, which apply to the appearance of DNA
- Logic processors, such as the brain
- Sensors and responders, such as the immune system and the five senses
- System coordination, as with organs and homeostasis
- Electrical networks, such as biological nerve systems
- Electronic measuring devices, such as those used in fossil-dating equipment

To begin with, if you were to open a book and saw letters of the alphabet splattered randomly all over each page, with letters backwards, upside down, and not in rows of any kind, you might assume that a printer had gone haywire. Without words or sentences, such a book would of course make no sense. Without order or structure, it would have no value whatsoever.

However, if you were to open a play written by Shakespeare, you would immediately recognize that the words were put in place by a deliberate source of intelligence. In fact, you could open the play to any page, and within the first three or four words, it would be apparent that everything on that page had been intelligently written. It would be intuitively obvious that it could not have been accidentally produced by randomly scattering letters on paper.

Similarly, if you were to read a symphony by Mozart, you would immediately recognize that everything was arranged with deliberate intent and was not an accidental combination of notes. Every single section of the symphony would indicate intentional, deliberate construction.

This analogy also pertains to computer programming. If you were to take a string of 0's and 1's, and at the very beginning they translated into, "Four score," you would have a strong suspicion that you were not reading something random. If it continued with the words, "and seven years ago," you would know for sure that it wasn't random and wouldn't be surprised when it continued with, "our forefathers set forth upon this continent a great nation…"

With each of the above examples, a source of intelligence is immediately obvious, whether it is the work of Shakespeare, Mozart, or a computer programmer. Therefore, what should we think when we consider DNA, which is the incredibly small yet massively complex building block that creates all life on Earth? It should be obvious that DNA exhibits intentional design, purpose, and function, and was in no way accidental.

If we were to print out just a single copy of human DNA, it would result in a book 750,000 pages long. To help visualize this, picture in your mind a book as tall as a 25-story building. That 25-story book would contain just one copy of DNA! Furthermore, on every one of those 750,000 pages, each segment of genetic code would reveal intricate and complex details about every human being on the planet. Some pages might contain anatomical instructions for the knee, others the brain, the stomach, etc. There would be genetic code for individual eye color, hair color, skin color, as well as for physical build, and so on. It's all there, laid out in extremely long lines of amines that form perfect genes that form perfect chromosomes that form all living organisms exactly as intended.

As for big bangs, that entire discussion is basically just a big diversion. Big bangs could not produce DNA. For that matter, big bangs could not even produce basic cellular components. Lastly, big bangs couldn't have anything to do with the design of the thousands upon thousands of species that work together to form the unimaginably complex ecosystem of this miraculous blue-green planet we call home.

Note from the author:

Many arguments presented in this book, from the smallest living cell on up, will be intentionally ignored by those whose understanding is based on the evolution story's viewpoint of the known world. But that is to be expected, since creation has been effectively eliminated from public school classrooms – because it specifically threatens the status quo. So we can logically believe if educators are dishonest in small things, they can't be trusted in xxx-extra-large matters either.

Visit EvolutionIsBS.com for more information.

Appendices

APPENDIX 1

The following proofs for Chapter Two are provided for anyone interested in confirming the accuracy of the figures presented.

For each of the three examples given, there are only two calculations necessary. The first simply determines how many keystrokes the monkeys would make. The second determines how long a run of letters would be produced in the proper order.

For the lone monkey example, there are 60 seconds in a minute, times 60 minutes in an hour, times 365.25 days in a year (averaging for leap years), times 70 years, times two keystrokes per second, which calculates out to about 9.2×10^7 *keystrokes taken by the lone monkey altogether during its lifetime.*

The next calculation for the lone monkey will determine how much of "ocimixiiixiixicicxcmooomiixixiiixicc" would be accidentally typed with those 9.2×10^7 *key-strokes, as follows: the odds of getting the first letter correct on any given keystroke are 1 in 26. The odds of consecutively getting the second letter correct are 1 in 26 times 26, which equals 676. Jumping ahead a few letters, calculations show that the odds of getting the first letter through the 6th letter correct are 1 in* 26^6*, which works out to approximately* 3.09×10^8*, which is more than triple the* 9.2×10^7 *keystrokes that the monkey made. That means that the very best the lone monkey could achieve over its 70-year lifetime would be the first five letters strung together in the desired order.*

In the example using ten monkeys, there are 60 seconds in a minute, times 60 minutes in an hour, times 10 hours per day, times 365.25 days in a year, times 70 years, times four keystrokes per second per monkey, which calculates out to roughly 1.84×10^{10} *keystrokes taken altogether by the ten monkeys.*

The next calculation for the ten monkeys will determine how much of "ocimixiiixiixicicxcmooomiixixiiixicc" would be accidentally typed *with those 1.84 x 10^{10} keystrokes, as follows: The odds of getting the first letter correct on any given keystroke are 1 in 26. The odds of consecutively getting the second letter correct are 1 in 26 times 26, which equals 676. Jumping ahead several letters, it calculates out that the odds of getting the first letter through the 8th letter correct are 1 in 26^8, which comes to about 2.09 x 10^{11}, which is more than ten times the 1.84 x 10^{10}keystrokes that the ten monkeys made. That means that the very best the ten monkeys could achieve as a group over their 70-year lifetime would be the first seven letters of the randomly selected "ocimixiiixiixicicxcmooomiixixiiixicc."*

In the third example regarding the million monkeys trying to type out "ocimixiiixiixicicxcmooomiixixiiixicc," there are 60 seconds in a minute, times 60 minutes in an hour, times 24 hours in a day, times 365.25 days in a year, times 10^{11}(one hundred billion) years, times 10^6 monkeys, times 100 keystrokes per second per monkey, which calculates out to roughly 3.16 x 10^{26} keystrokes taken altogether.

The final calculation determines how much of "ocimixiiixiixicicxcmooomiixixiiixicc" the monkeys would type with those 3.16 x 10^{26} keystrokes. The odds of any one of the monkeys getting the first letter correct on any particular keystroke are 1 in 26. The odds of consecutively getting the second letter correct are 1 in 26 times 26, which equals 676. Jumping ahead a little, it just so happens that the odds of any of the monkeys getting the first letter through the19th letter correct are one in 26^{19}, which calculates out to about 7.66 x 10^{26} keystrokes, which is more than double the 3.16 x 10^{26} keystrokes that the monkeys made. That means that the very best the monkeys together as a group could possibly achieve would be only the first 18 letters of "ocimixiiixiixicicxcmooomiixixiiixicc" strung together in the proper order over the course of 100 billion years.

APPENDIX 2

The following proofs are provided for those who are interested in confirming the accuracy of the figures presented in Chapter Six:

For the example with eight codons, the odds of being correct for each codon position, respectively from left to right, are one in 32, one in 16, one in 32, one in 16, one in 16, one in 32, one in 32, and one in 21.3333. Multiply those numbers times one another and it calculates to 91,625,967,549 codons needed in all.

For the example with the 192 codons (576 ATCG's), multiply as in the example above, but with significantly more codons. Each of the 192 codons used has odds of either one in 10.667, or 16 to one, or 21.333 to one, or 32 to one. Once the numbers have been crunched and the dust settles, the result is 7.7202×10^{233}*, which is the following number, give or take a very tiny percentage:*

772,020,010,011,010,110,000,101,110,100,011,101,000,110,100, 001,100,101,011,101,110,010,000,000,100,000,011,011,100,110,100, 101,101,110,011,001,010,111,010,001,100,101,011,001,010,110,111, 000,100,000,001,000,000,111,010,001,110,111,011,001,010,110,111, 001,110,100,011,110,010,010,000,001,110,011,011,010,010,111.1000.

APPENDIX 2

APPENDIX 3

The following proofs are provided for those who are interested in confirming the accuracy of the figures presented in Chapter Seven:

For those interested in how the first meteor story analysis works, there are two calculations involved. The first calculation determines the total number of amines deposited into the earth's atmosphere, and the second determines the length of the longest amine run produced during the one million year period.

First, we have 10,000 meteors per day, times 365.25 days per year, times one million years, times 10,000 amines per meteor, which yields 3.65×10^{16} *amines, which divided by three produces* 1.22×10^{16} *codons.*

Next, the odds against accidentally having the first codon position correct are one in 20. The odds of also getting the second position correct are one in 20 times 20 (or 20^2*), which equals 400. Jumping ahead, to get 13 codons lined up correctly would take* 20^{13} *codons, which when multiplied out comes to* 8.19×10^{16} *codons, which is several times more than the* 1.22×10^{16} *codons available. So the odds are that this set of meteors would achieve no more than 12 codons (which is 36 amines), since 13 codons would be virtually impossible.*

For those interested in how the larger meteor story analysis works, there are three calculations necessary. The first determines the total number of meteors that would enter the earth's atmosphere. The second determines how many total amines those meteors would contain. The third determines the length of the longest correctly ordered string of amines that would exist at any one time during that period.

For the first calculation, we know from Appendix One that there are 3.16×10^7 *seconds in a year, and multiplying that by* 3×10^9 *(three billion) years works out to roughly* 9.48×10^{16} *meteors in all.*

Second, there are 10^7 (ten million) amines per meteor, times 9.48×10^{16} meteors, which equals 9.48×10^{23} amines total, which when divided by three produces 3.16×10^{23} codons .

Next, the odds against accidentally having the first codon position correct are one in 20. The odds of also getting the second position correct are one in 20 times 20 (or 20^2), which equals 400. Jumping ahead, to get 19 codons lined up correctly would take 20^{19} codons, which when multiplied out comes to approximately 5.24×10^{24} codons, which is more than the 3.16×10^{23} codon combinations available. So the odds are that this enormous set of amine-rich meteors would achieve no more than 18 codons (which is 54 amines), since 19 codons would be virtually impossible.

For those interested in how the ocean story analysis works, there are two calculations necessary. The first determines the total number of amines that are contained therein. The second determines the length of the longest correctly ordered string of amines that would exist.

The ocean is 1,000 miles long, times 1,000 miles wide, times 10 miles deep, which equals 10,000,000, or 10^7 cubic miles. A cubic mile is 5,280 feet wide, times 5,280 feet long, times 5,280 feet deep, which equals 1.47×10^{11} cubic feet. So, 10^7 cubic miles times 1.47×10^{11} cubic feet per mile equals 1.47×10^{18} total cubic feet. There are approximately 2.83×10^{28} amines per cubic foot; therefore, the total number of amines in the subject ocean is 2.83×10^{28} times 1.47×10^{18}, which equals 4.17×10^{46} amines, which divided equals 1.39×10^{46} codons .

Next, the odds against accidentally having the first codon position correct are one in 20. The odds of also getting the second position correct are one in 20 times 20 (or 20^2), which equals 400. Jumping ahead, to get 36 codons lined up correctly would take 20^{36} codons, which when multiplied out comes to approximately 6.87×10^{46} codons, which is more than the 1.39×10^{46} codon combinations available. So the probabilities are that this particular ocean would achieve no more *than 35 codons (which is 105 amines), since 36 codons would have to overcome impossible odds.*

For the planet Earth analysis, its volume is 4/3 π r^3, which is 4/3 π (3,963.2 miles)^3, which is 2.61 x 10^{11} cubic miles. A cubic mile is 1.47 x 10^{11} cubic feet. So, 2.61 x 10^{11} cubic miles times 1.47 x 10^{11} cubic feet per cubic mile, equals 3.84 x 10^{22} total cubic feet. There are approximately 2.83 x 10^{28} amines per cubic foot. Therefore, the total number of amines in the Earth story is 2.83 x 10^{28} times 3.84 x 10^{22}, which equals 1.09 x 10^{51} amines.

For the second part of the calculation, the amines shuffle ten times per second for one hundred billion years, which would be 4.17 x 10^{46} seconds per year, times 10^{11} (one hundred billion) years, times ten shuffles per second, times 1.09 x 10^{51} amines, which works out to 3.44 x 10^{70} amines aligned randomly, which divided by three yields 1.15 x 10^{70} codons.

Next, the odds against accidentally having the first codon position correct are one in 20. The odds of also getting the second position correct are one in 20 times 20 (or 20^2), which equals 400. Jumping ahead, to get 54 codons lined up correctly would take 20^{54} codons, which when multiplied out comes to approximately 1.80 x 10^{70} codons, which is more than the 1.15 x 10^{70} codon combinations available. So the probabilities are that this amine-planet story would achieve no more than 53 codons (which is 159 amines), since 54 codons would have to overcome impossible odds.

APPENDIX 4

CAGGCGTTCGGACGTGAGAAATACATCAATGGATAGATCTCGAGAGAAA
GAACATACGCCCTCAAGAACGTGCGTTCTCAAGAACTGCCGTTCTCCCT
AGAGAACGGACGCCCGACCTAGCTCAAGAACGAGCGCCAGAACTCACTA
GCGTTCTCCCGAGCGTACGCCCGCAAGTAAGAACGTGCGCCCGGCCTCA
CGGACGCCCTAGAGAACGTACGCCCTCAAGAACGGCCTCAAGAACGAGC
GCCAGAACGACCGCGCTAGCGACCGGCCGCAAGTGAATCAAGGATAGAT
GAAGAAAGAACCGCCGCCAGAACGGACGACCTCGCGCCAGAACGGACGC
CCGACCTAGCGCAAGAACGGACGTTCTCTAGAACAGCAGAACTATCGAC
CGGCCGCAAGAACTCCCGTGCTCACGTTAGAACTGCCGTTCTCCAGTAA
GAACAGCAGAACGCTCGTTAGAACGACCTCTCGACCTGCAGTAAGAACG
ACCGTGCGCAAGAACGATCGTTCGTCCGCCAGAACGACCGCTCGACCGG
CCGTGAGAACTCCCGTGCTCACGTTAGAACTGCCGTTCTCCAGTGAGAA
CAGCCGCGAGAACTGCCGCCAGAACGTACGTTCTCGCGCCCGCAAGAAC
GTCCGCCAGTAAGAACTGCCGCCAGAACTCTCGTTCTCCCGTACGCAAG
AACTAGCGCCCGGGCGTTCGGCCGATCGCCAGTAAGAACGAGCGCCCGA
TCGACCTCCCTATCGCCAGAACAGCAGAACTATCGACCGGCCGCAAGTA
AGAACAGCAGAACGCTCGTTAGAACTCCCGTGCTCACGTTAGAACTCAC
GGACGCCAGAACACGCGACCTCACGGACGCCCTAGATGGAGAACGCGCG
TTCTAGAGAACGTCCTGCAGAACACGCGACCTCACGGACGCCCTAGAGA
ACGGCCTATAGAACGCTCTAGCGCCCGACCTCACGCCCTAGAGAACTCA
CGGACGACCGTGAGAACAGCAGTGAATCAAGGATAGATGCAGAAAGAAC
AACCGTGCGCAAGAACGTGCGTTCTCTAGAACAGCAGAACGGACGACCT
CGCGCCAGAACTCACGTTCGTACGCAAGAACTGCCGTTCTCCAGAACGA
GCGCCCGCGCGTTCTAGCGCCAGAACGGCCTCAAGAACGATCGACCGTC
CGCCAGAACTCACGTTAGAACTAACGACCTATCTATAGTAAGAACTCAC
GGACGACCTCAAGTAAGAACTCTCGGACGCCCGTGAGAACGGCCTCAAG
AACGGCCTATAGAACGATCGTTCGTCCGCCAGAACTCACGTTAGAACTA
ACGACCTATCTATAGTAAGAACTGCCGCCAGAACGTCCGGCCGCTCGGA
CTCAAGAACGAGCGCCCGTACGGCCGCCCTCGCGCCAGTGAATCAAGG.

CRACKING THE CODES

This section is provided for readers who would prefer some hints about this book's hidden messages. Every code is fairly tricky – they are codes after all!

Chapter One (page 23): First paragraph only, start with the "G" in "Generally" and read straight down along left margin.

Chapter Two: Change "ooimcmooomoixixiiixiicmmooooimciix-micmimo" into Roman numeral equivalents where each o is 0, each i is 1, each x is 10, each c is 100, and each m is 1000. Then use a binary-to-text converter.

Chapter Three: Enter each string of 0's and 1's on pages 35 and 36 into a binary-to-text convertor.

Chapter Five: First message, middle of page 46: in the small sample string of 0's and 1's and periods, change into Morse code: 0 is dot, 1 is dash, period is space. Second message: final paragraph of chapter, read down left-hand side and then down right-hand side.

Chapter Six (near top of page 53): Enter 0's and 1's after 772,020 into binary-to-text converter.

Chapter Eight (page 68, top paragraph, last sentence): Use the first letter of each word in the list, but don't use the after-hyphen half of any hyphenated word.

Chapter 10 (bottom page 77, top page 78): First letter of each first word in organ list spells three-word message.

Chapter 13(bottom page 95, top page 96): Use the first letter of first word of each line item in immune system list.

Chapter 14 (page 103): First letter of each first word in anatomical list spells three-word message.

Chapter 15 (page 109): Start with the "J" in "Just" and read straight down along the left margin for the first three paragraphs.

Chapter 18 (page 135): In "Note from the author" (last paragraph), take first letter of first word and of every third word thereafter (first letter of 1st word, 4th word, 7th word, etc.)

Chapter 20 (bottom page 147, top of page 148): Enter string of 0's and 1's after 194,267,000 into binary-to-text converter.

Conclusion: Final paragraph of chapter, read down left-hand side and then down right-hand side.

Appendix 4 (page 181): This code and those similar in Chapters Six and 20 are reserved for dedicated code-breakers.

Cracking the Codes: The last paragraph on page 185 contains what might be the toughest code of them all, and could require the assistance of an experienced code-breaker. (Hint: message involves "cross words.")

Closing Words (page 187): Start with the "H" in "Hidden," and read straight down along the left margin, full page. Forms a five-word message, provide your own two spaces between the first three words. *(Page 188):* Start with the "R" in "Rational" and read straight down along the left margin, full page.

Back Cover: Start with the "P" in "Perhaps" and read straight down along left-hand side; then read down right-hand side, last letter of last word in each line for eight lines, ignoring punctuation marks.

(And there could be some additional "bonus codes" of various sorts hidden here and there...)

Many of the hidden messages might not be terribly difficult to decode with the hints provided. But imagine how hard it would be to go from encoding only 576 bits of information (see the hidden code in Chapter Six) to encoding the *3 billion* amines that comprise the genetic code for a human being. It would be more than *4 million* times as much work simply to encode it, let alone design it!

Code-breaking concerns the exposure of special hidden messages. Many a decipherer enjoys spending hours on end looking for clues. However, some people choose to avoid involvement in possibly brain-numbing analysis. (Properly choosing your decoding mode will be helpful in picking up some logic, which gets you more logic.) Growing in science does not mean you should be blindly sticking up for one side. No, you should be requiring well-rounded reviews backing up science, since some ideologies, even if well-liked previously, can possibly be disproven.

CLOSING WORDS

No pleasure is comparable to the standing upon the vantage ground of the Truth.
– Sir Francis Bacon

Several messages have been encoded throughout this book. Hidden in plain sight (including one on this page), could these encryptions have happened simply by random coincidence? If not, then this book might end up getting banned from public schools, because atheism is now the only belief system allowed. Regulations have been enacted due to some very controversial interpretations of the First Amendment, which says Congress should not establish a religion. But it does not say anywhere that educators cannot teach anything regarding religion, just that a national religion cannot be put in place.

As a result, we must consider the far-reaching implications. Just a single line or two of hidden code could not have become encrypted accidentally in the various pages of this book, so what situation could cause three billion bits of *genetic* code to all line up perfectly, with each bit accidentally positioned in the precise spot required to create life?

You can't have it both ways in these types of considerations. Could this book's hidden messages *and* genetic code each have happened accidentally? The one alternative is that they both were rationally and intentionally constructed. If the messages hidden inside this book (and on its covers) occurred by design, it would seem certain that highly complicated, extensive genetic code had to have occurred by design also.

In time, the truth always becomes obvious. It is inevitable. Rational thought combines with newly attained information, eventually building upon previous ideas sufficiently enough to verify prior beliefs – *or* to pave the way for the consideration of entirely new concepts. Original ideas are accompanied by high levels of anticipation, and might typically generate a great deal of attention and excitement. Some concepts, such as the microchip, the light bulb, and the horseless carriage all ultimately entailed inventions that were new to the world – and were confirmed for one and all to see. Nonetheless, other ideas were centered around notions that later proved to be completely false.

As such, we have come full-circle with the evolution story. There was much fanfare when the concept was first introduced way back in 1859, and ever since then it has been very intensely evaluated, with much argumentation for and against. However, now that the internet exists, the near-stranglehold on information that universities used to enjoy no longer exists. Ideas that only yesterday seemed to be solidly scientific are suddenly vulnerable to wide exposure. Therefore, it is simply a matter of steadfastly working to inform those who will listen, and ultimately the truth of the matter will be illustrated for all to see.

All things deserve scrutiny, including the story of evolution. There will be a time in the not-too-distant future when people will wonder how anything so unscientific, and so illogical, could ever have been widely accepted as fact. Unfortunately, there was not enough information put forth in the past. But today, modern technological availability allows for widespread computer usage, yielding access to contrary information. The matter then becomes one of scientific integrity. How long can evolutionists deny that nearly everything in their favorite story is basically impossible, especially when the odds involved are infinitely against them?

Made in the USA
Monee, IL
04 November 2026

67386596R00105